NUREG-1880

ATHEANA User's Guide

Final Report

Manuscript Completed: May 2007
Date Published: June 2007

Prepared by:

John Forester, Sandia National Laboratories
Alan Kolaczkowski, Science Applications International Corp.
Susan Cooper, U.S. Nuclear Regulatory Commission
Dennis Bley, Buttonwood Consulting, Inc.
Erasmia Lois, U.S. Nuclear Regulatory Commission

Erasmia Lois: NRC Project Manager

Prepared for:
Division of Risk Assessment and Special Projects
Office of Nuclear Regulatory Research
U.S. Nuclear Regulatory Commission
Washington, DC 20555-0001

ABSTRACT

This manuscript provides a user's guide for the human reliability analysis (HRA) method known as "A Technique for Human Event Analysis" (ATHEANA), which the U.S. Nuclear Regulatory Commission (NRC) documented in NUREG-1624, Rev. 1, dated May 2000. As the first publication of its kind, this user's guide describes both the quantitative and qualitative ATHEANA analysis approaches, fully describing the revised quantification approach and presenting a simpler description of the other ATHEANA elements needed to perform an HRA as part of a probabilistic risk assessment (PRA). Toward that end, this user's guide strives to present the steps for applying ATHEANA in a straightforward and succinct manner, so that HRA experts can easily and effectively apply the technique. Consequently, although the authors relied on NUREG-1624 as a primary resource for its development, the NRC is publishing this user's guide as a standalone document, such that it can be used by analysts to apply the ATHEANA technique without the need to use NUREG-1624.

PAPERWORK REDUCTION ACT STATEMENT

This user's guide contains information collection requirements that are subject to the Paperwork Reduction Act of 1995 (44 U.S.C. 3501 et seq.). These Office of Management and Budget (OMB) approved these information collections under 10 CFR Part 50, approval number 3501-0011.

Public Protection Notification

The NRC may neither conduct nor sponsor, and a person is not required to respond to, a request for information or an information collection requirement unless the requesting document displays a currently valid OMB control number.

FOREWORD

This user's guide supports the human reliability analysis (HRA) method known as "A Technique for Human Event Analysis" (ATHEANA), which the U.S. Nuclear Regulatory Commission (NRC) documented in NUREG-1624.[1] ATHEANA is a method for identifying plausible error-likely situations and potential error-forcing contexts that may result in human failure to correctly perform an action, and for estimating human error probabilities (HEPs) for the human events modeled in probabilistic risk assessments (PRAs).

The objective of this user's guide is to provide step-by-step guidance on how to use ATHEANA when performing HRA. As such, it describes the steps for applying ATHEANA in a straightforward and succinct manner, so that HRA experts can easily and effectively apply the technique. In addition, this user's guide documents the ATHEANA quantification approach which was developed after NUREG-1624 was published. It also incorporates lessons learned from the use of the ATHEANA qualitative and quantitative approaches in NRC-sponsored PRAs and HRAs. This user's guide aims to provide a clearer understanding of its advantages, as well as a guidance on the types of regulatory applications for which ATHEANA would be most suitable and would yield the greatest benefit.

NUREG-1624 provides guidance for two analysis approaches, namely "retrospective analysis" — a process used to analyze important historical events from a human performance perspective, and "prospective analysis" — a process used to analyze human failure events modeled in PRAs. By contrast, this user's guide deals strictly with prospective analysis, consistent with other commonly used HRA methods.

Although the authors of this guide relied on NUREG-1624 as a primary resource for its development, the NRC is publishing this user's guide as a standalone document, such that it can be used by analysts to apply the ATHEANA technique without the need to use NUREG-1624. While NUREG-1624 contains additional useful information that supports this user's guide, in many cases, that additional information may be better used after analysts are more experienced with the basic method as presented in this guide.

Brian W. Sheron, Director
Office of Nuclear Regulatory Research
U.S. Nuclear Regulatory Commission

---

[1]   All references to NUREG-1624 apply to Rev. 1, "Technical Basis and Implementation Guidelines for A Technique for Human Event Analysis (ATHEANA)," dated May 2000.

CONTENTS

Figures

Tables

ACKNOWLEDGMENTS

The authors would like to thank the following individuals who served as peer reviewers for this document: Harold Blackman, Yung Hsien J. Chang, Vinh Dang, Jeff Julius, Ken Kiper, Ali Mosleh, and Oliver Straeter. The comments and suggestions they provided were extremely beneficial. In addition, the authors would like to acknowledge the contributions to the ATHEANA project by Nathan Siu and Gareth Parry. Special thanks goes to Gareth Parry who reviewed and provided feedback throughout the development of this document. Finally, the authors acknowledge the contributions from Ann Ramey-Smith and John Wreathall to the ATHEANA method development.

ABBREVIATIONS

AFW	auxiliary feedwater
AOP	abnormal operating procedure
ASME	American Society of Mechanical Engineers
ATHEANA	A Technique for Human Event Analysis
BWR	boiling-water reactor
CESA	Commission Errors Search and Assessment
CFR	*Code of Federal Regulations*
DC	direct current
EFC	error-forcing context
EOC	error of commission
EOO	error of omission
EOP	emergency operating procedure
EPRI	Electric Power Research Institute
FSAR	final safety analysis report
HAZOP	HAZard and OPerability study
HEP	human error probability
HFE	human failure event
HRA	human reliability analysis
HSI	human-system interface
LOCA	loss-of-coolant accident
MFW	main feedwater
NASA	National Aeronautics and Space Administration
NRC	U.S. Nuclear Regulatory Commission
PORV	pressure-operated relief valve
PRA	probabilistic risk assessment
PSA	probabilistic safety assessment
PSF	performance-shaping factor
PTS	pressurized thermal shock
PWR	pressurized-water reactor
RCS	reactor coolant system
RPS	reactor protection system

SAMG	severe accident management guideline
SG	steam generator
SHARP1	A Revised Systematic Human Action Reliability Procedure
SI	safety injection
SRV	safety relief valve
SS	shift supervisor
T-H	thermal-hydraulic
UA	unsafe action

GLOSSARY

Aleatory Uncertainty: Random variability in any of the factors that lead to variability in the results. Aleatory uncertainty (1) is (or is modeled as) irreducible, or (2) is observable (i.e., repeated trials yield different results), or (3) exists when repeated trials of an idealized thought experiment will lead to a distribution of outcomes for the variable (this distribution is a measure of the aleatory uncertainties in the variable).

Availability Heuristic: The tendency of individuals to base interpretations or judgements on the ease with which relevant information can be recalled or with which relevant instances or occurrences can be imagined. Availability can be influenced by factors such as the recency and primacy of the individual's own experiences.

Deviation Scenario: A plausible deviation from the nominal conditions or plant evolutions normally assumed for the probabilistic risk assessment (PRA) sequence of interest (the nominal scenario), which might cause problems or lead to misunderstandings for the operating crews.

Epistemic Uncertainty: When the state of knowledge about the effects of specific factors is less than perfect. With epistemic uncertainty, (1) we are dealing with uncertainties in a deterministic variable for which the true value is unknown, or (2) repeated trials of a thought experiment involving the variable will result in a single outcome that is the true value of the variable, or (3) the uncertainty is reducible (at least in principle).

Error-Forcing Context (EFC): The situation that arises when particular combinations of *performance shaping factors* and *plant conditions* create an environment in which unsafe actions are more likely to occur.

Error of Commission (EOC): A *human failure event* resulting from an overt, unsafe action, that, when taken, leads to a change in plant configuration with the consequence of a degraded plant state. Examples include terminating running safety-injection pumps, closing valves, and blocking automatic initiation signals.

Error of Omission (EOO): A *human failure event* resulting from a failure to take a required action, that leads to an unchanged or inappropriately changed plant configuration with the consequence of a degraded plant state. Examples include failures to initiate the standby liquid control system, start auxiliary feedwater equipment, or block automatic depressurization system signals.

Frequency Bias/Effects: Frequently occurring events are often recalled more easily than rare events. This can lead to a tendency for people to interpret incoming information about an event in terms of events that occur frequently, rather than infrequently occurring or unlikely events.

Heuristic: A way of mentally taking a shortcut in recognizing a situation. Heuristics normally allow people to quickly select the most plausible choices first, followed by the less plausible choices.

Human Error: In the PRA community, the term "human error" has often been used to refer to human-caused failures of systems or components. However, in the behavioral sciences, the same term is often used to describe the underlying psychological failures that may cause the human action that fails the equipment. Therefore, in ATHEANA, the term "human error" is only used in a very general way, with the terms *human failure event* and *unsafe action* being used to describe more specific aspects of human errors.

Human Failure Event (HFE): A basic event that is modeled in the logic models of a PRA (event and fault trees), and that represents a failure of a function, system, or component that is the result of one or more unsafe actions. A human failure event reflects the PRA system's modeling perspective.

Information Processing Model: A general description of the range of human cognitive activities required to respond to incoming information. The model used in ATHEANA considers activities in response to abnormalities as involving the four steps of (1) monitoring/detection, (2) situation assessment, (3) response planning, and (4) response implementation.

Mental Model: Mental representations that integrate a person's understanding of how systems and plants work. A mental model enables a person to mentally simulate plant and system performance in order to predict or anticipate plant and equipment behavior.

Monitoring/Detection: The activities involved in extracting information from the environment. Monitoring is checking the state of the plant to determine whether the systems are operating correctly. Detection, in this context, refers to the operator becoming aware that an abnormality exists.

Nominal Scenario: The evolution of a PRA scenario that is nominally expected, or is at least a good representative case, given the level of detail provided by the PRA model or other modeling framework being used. A description of the nominal scenario provides a basic understanding of the progression of events associated with the scenario as defined in the PRA, as well as denoting key characteristics that add to the understanding of the scenario, such as the expected timing of significant plant status changes and the expected trajectories, over time, of key parameters. The nominal scenario is that typically modeled in the PRA.

Performance Shaping Factors (PSFs): A set of influences on the performance of an operating crew resulting from the human-related characteristics of the plant, the crew, and the individual operators. Example characteristics include procedures, training, and human-factors aspects of the displays and control facilities of the plant.

Plant Conditions: The plant state defined by combinations of its physical properties and equipment conditions, including the measurement of parameters.

Primacy Bias/Effects: The tendency for people to assign greater significance to the data they first see (and from which they may draw conclusions) than to later data. When judgments or decisions are required, initial information is sometimes more easily recalled than subsequent information.

Probabilistic Risk Assessment (PRA): For a nuclear power plant, a PRA is an analytical process that describes and quantifies the potential risk (associated with the design, operation, and maintenance of the plant) to the health and safety of the public.

PRA Model: A logic model that generally consist of event trees, fault trees and other analytical tools, and is constructed to identify the scenarios that lead to unacceptable plant accident conditions, such as core damage. The model is used to estimate the frequencies of the scenarios by converting the logic model into a probability model. To achieve this aim, estimates must be obtained for the probabilities of each event in the model, including HFEs.

PRA Scenario/Sequence: The scenario developed in terms of that readily discernible from the PRA or other risk-related framework that is providing an initial model of the sequence of events involving the human action(s) of interest. The PRA scenario is based on the minimum descriptions of plant state required to develop the PRA model and define appropriate HFEs. Examples of scenario definition elements include the initiating event, operating mode, decay heat level (for shutdown PRAs), and function/system/component status or configuration. The level of detail to which scenarios are defined can vary and include the functional level, system level, and component level.

Recency Bias/Effects: Events that happened recently are recalled more easily than events that occurred a long time ago. In attempting to understand incoming information about an event, people tend to interpret the information in terms of events that have happened recently, rather than relevant events that occurred in the more distant past.

Representativeness Heuristic: The tendency to misinterpret an event because it resembles a "classic event" which was important in past experience or training, or because there is a high degree of similarity between the past event and the evidence examined so far.

Response Implementation: Taking the specific control actions required to perform a task, in accordance with *response planning*. Response implementation may involve taking discreet actions (e.g., flipping a switch) or it may involve continuous control activity (e.g., controlling the steam generator level). It may be performed by a single person, or it may require communication and coordination among multiple individuals.

Response Planning: Deciding on a course of action, given a particular *situation model*. In general, response planning involves identifying plant-state goals, generating one or more alternative response plans, evaluating the response plans, and selecting the response plan that best meets the goals identified.

Rules: Rules are the guidance operators follow in carrying out activities in the plant. Rules can be either formal or informal in nature. *Formal rules* are specific written instructions and requirements provided to operators and authorized for use by plant management. *Informal rules* sources include training programs, discussions among operators, experience, and past practices.

Salience Bias: The tendency to give closer attention or to weight more heavily information or indications that are more prominent, (e.g., the most visible, the loudest, or the most "compelling" instrument displays.)

Situation Assessment: Situation assessment involves developing and updating a mental representation of the factors known, or thought to be affecting the plant state, at a given point in time. The mental representation resulting from situation assessment is referred to as a situation model.

Situation Model: A mental representation of the current plant condition, and the factors thought to be affecting the plant state resulting from the operators' situation assessment. The situation model is created by an interpretation of operational data in light of the operator's mental model. (An operator's situation model is usually updated constantly as new information is received; failure to update a situation model to incorporate new information is an error mechanism).

Unsafe Action (UA): Actions inappropriately taken, or not taken when needed, by plant personnel that result in a degraded plant safety condition. In ATHEANA, the potential for multiple UAs contributing to a particular HFE is considered.

1. INTRODUCTION

1.1 Motivation for ATHEANA

"A Technique for Human Event Analysis," or ATHEANA, is a human reliability analysis (HRA) methodology designed to support the understanding and quantification of human failure events (HFEs) in nuclear power plants. On the basis of reviews of operating experience in technically challenging domains such as nuclear power plants, a key observation that drives the ATHEANA approach is that HFEs that contribute to equipment damage or other severe consequences, and that involve highly trained staff using considerable procedure guidance, do not usually occur randomly or as a result of simple inadvertent behavior such as missing a procedure step or failing to notice certain indications because they are on a back panel. Instead, such HFEs occur when the operators are placed in an unfamiliar situation where their training and procedures are inadequate or do not apply, or when some other unusual set of circumstances exists.

In such situations, even highly trained staff often make incorrect assessments regarding the status of the system being monitored or controlled, and subsequent human actions may not be beneficial or may even be detrimental. The following examples are representative of the numerous instances of HFEs that occur in such situations:

- In the Three-Mile Island accident, several equipment failures occurred (all auxiliary feedwater unavailable at least for a short time, with a partially stuck-open power-operated relief valve on the pressurizer), which together might typically be considered to have a low probability. The resulting plant response and related indications led the operating crew to believe that the reactor coolant system was solid (full). As a result, the crew inappropriately stopped all safety injection, and persisted in this response despite indications that the plant situation was becoming seriously degraded.

- In the Chernobyl accident, the operators implemented a series of bad decisions, which might typically be considered too improbable. Nonetheless, those decisions placed the plant in an inherently unsafe state and ultimately led to the accident. Moreover, the operators initially disbelieved that the accident had even occurred.

- In the Air Florida 737 aircraft crash into the 14th Street Bridge in Washington, DC, the pilots exacerbated the icing conditions on the wings before takeoff. Further, they did not comprehend the significance of the instrumentation and throttling anomalies, which were influenced by a faulty, high-thrust indication caused by ice on a pressure probe. Finally, the pilots then failed to increase thrust during the initial climb because they failed to understand the seriousness of the situation.

A review of these and other serious events leads to the following conclusions about common characteristics that, when sufficiently strong, cause human errors that lead to significant consequences:

- Plant or system behavior is outside the expected design range, such as when multiple or cascading equipment failures or unavailabilities occur.

- Plant or system behavior is not understood because, for instance, the behavior is outside the operators' expectations based on their training and experience.

- Indications of the plant or system state or behavior are not recognized or are even misleading because of instrumentation failures or other anomalies.

- Prepared plans or procedures are not applicable nor helpful because the actual conditions or the evolution of the event is beyond that envisioned when the guidance was developed.

Fortunately, situations with these characteristics do not often occur. However, when they do, the risk of a serious error can be quite high.

> ATHEANA is an HRA methodology designed to (1) identify plausible error-likely situations and potential error-forcing contexts (EFCs), and (2) produce estimated human error probabilities (HEPs) in risk assessments.

ATHEANA is an HRA methodology designed to search for situations with one or more of the above characteristics, and estimate the probability of making an error in such situations for use in a probabilistic risk or safety assessment (PRA/PSA). Such situations are said to have an error-forcing context (EFC) in ATHEANA terminology. In addition, because situations with a strong EFC may not always be likely, ATHEANA provides guidance for evaluating behavior in the more nominal case that is typically modeled in a PRA.

1.2   Purpose

The purpose of this user's guide is to provide PRA and HRA analysts with the following assets:

- better understanding of the ATHEANA process

- suggestions regarding when to apply ATHEANA, along with a discussion of its advantages

- step-by-step guidance on how to apply ATHEANA, including its approach for quantifying HFEs [i.e., estimating human error probabilities (HEPs)].

In order to achieve its purpose, this user's guide presents a simplified version of the multi-step analysis process covered in NUREG-1624 [Ref. 1][2]. By focusing on the more essential elements of ATHEANA, this user's guide provides sufficient guidance for practitioners to perform an HRA using ATHEANA, even though they may be first-time users of this method. Later, as a user becomes more proficient, some of the more extensive guidance in NUREG-1624 can be utilized (i.e, when the analyst can be more appreciative of the nuances and details, including the behavioral sciences aspects, provided in that document).

The ATHEANA process can be used for retrospective analysis of actual events, as well as prospective analysis of hypothetical HFEs. This user's guide deals strictly with prospective analysis consistent with other commonly used HRA methods. In theory, ATHEANA can be used for pre-initiating human events and post-initiating human events in nuclear power activities, as well as for other-than-commercial nuclear power technologies. However, ATHEANA guidance and experience with its use are most mature for post-initiating human events in nuclear power plant applications. Therefore, this user's guide focuses on these uses.

---

[2]   All references to NUREG-1624 apply to Rev. 1, "Technical Basis and Implementation Guidelines for A Technique for Human Event Analysis (ATHEANA)," dated May 2000.

1.3   Background

NUREG-1624 comprehensively documents the human reliability analysis method known as "A Technique for Human Event Analysis" (ATHEANA), and includes the following aspects:

*   behavioral sciences background and underlying model for human performance, upon which ATHEANA is based

*   terminology used by ATHEANA

*   rationale for ATHEANA (based on operating experience)

*   detailed process to implement the method (except for the more recently developed quantification process contained in this user's guide)

Overall, NUREG-1624 provides guidance for two analysis approaches:

(1)   Retrospective Analysis: a process to analyze a past, actual event to determine its causes

(2)   Prospective Analysis: a process used in PRA to analyze a potential HFE, determine its possible causes, and estimate the corresponding HEP

By contrast, this user's guide addresses only prospective analysis. Since the publication of NUREG-1624, the ATHEANA prospective analysis has been used in support of PRAs and HRAs sponsored by the U.S. Nuclear Regulatory Commission (NRC). Most notably, the NRC's Office of Nuclear Regulatory Research used ATHEANA (in a somewhat abbreviated form) in conducting the HRA for the technical analyses related to pressurized thermal shock (PTS). This application, involving the combined efforts of nuclear industry and utility personnel, as well as NRC research staff and their contractors, contributed to the determination that the NRC can relax the current PTS Rule[3] without significant added risk from nuclear power plant operations extending beyond the typical original 40-year license.

This user's guide applies lessons learned from these applications of ATHEANA (including improvements to the quantification process), and provides those in the PRA/HRA community with a means for technology transfer as to how to use the methodology. ATHEANA is still evolving and will continue to do so as it is (hopefully) applied by a broader cadre of analysts for a broader range of applications. Feedback from broader use will improve ATHEANA as an HRA method, and will allow influences of ATHEANA to be (possibly) included in other HRA methods.

---

[3]   In this context, "the current PTS Rule" refers to Title 10, Section 50.61, of the *Code of Federal Regulations* (10 CFR 50.61), "Fracture Toughness Requirements for Protection Against Pressurized Thermal Shock Events," and the PTS screening limits therein.

1.4   Overall Scope and Key Distinctive Features

As an HRA methodology, much of the ATHEANA process is similar to that of other methods. It covers the major analytical steps in conducting an HRA, as addressed in the PRA Standard promulgated by the American Society of Mechanical Engineers (ASME) [Ref. 2], and as presented in the good practices for HRA described in NUREG-1792 [Ref. 3]. In particular, ATHEANA covers the following steps:

- Identify human actions to be assessed.

- Define HFEs pertinent to performing these human actions incorrectly.

- Determine the HEPs for the defined HFEs, including consideration of likely recovery actions.

In addition, ATHEANA includes the following formalized, structured, and documented processes, which comprise key distinctive features of the methodology:

- Identify operational vulnerabilities that could set up potential unsafe actions (UAs)[4] (e.g., procedure weaknesses and operator knowledge limitations and biases).

- Identify plausible deviations from nominal conditions or plant evolutions that might cause problems or misunderstandings.

- Identify important performance-shaping factors (PSFs) relevant to both nominal and deviation conditions.

- Identify other aleatory factors that could significantly affect the likelihood of the HFEs and their uncertainties (i.e., investigating a broad range of potential influences).

These features especially relate to searching for EFCs for the related HFEs, and determining the corresponding HEPs for inclusion in PRAs/HRAs.

---

[4]   Human failure events (HFEs) are basic events modeled in the PRA (event trees and fault trees) that represent function, system, or component failures resulting from one or more *unsafe actions*. In turn, unsafe actions (UAs) are actions inappropriately taken, or not taken when needed, by plant personnel that result in a degraded plant safety condition. In ATHEANA, the analysis considers the potential for multiple UAs contributing to a particular HFE.

2. OVERVIEW OF ATHEANA FOR PROSPECTIVE ANALYSIS

This section provides an overview of ATHEANA for performing a prospective HRA. Other subsections cover anticipated advantages and disadvantages of using the method, as well as criteria to consider in determining when to use ATHEANA.

2.1   Overview of the ATHEANA Methodology

Figure 2.1-1 provides an overview of the steps involved in the ATHEANA methodology, including those that may have been performed primarily by the modeler/analyst in traditional PRA (e.g., developing the overall scope of the analysis, including the PRA sequences, and defining the associated human actions of interest). The ATHEANA methodology covers these steps to emphasize the importance of including HRA-related thinking in the development of the PRA, and to describe the information that will be needed from these steps to support the HRA. Other ATHEANA steps address activities that are typically thought of as being more within the technical discipline of HRA. The documentation of an HRA using ATHEANA is not covered separately herein; rather, the reader is referred to References 2 and 3 for requirements and good practices related to documenting an HRA. However, Section 3 of this user's guide does discuss information that is particularly important to include in the documentation for an ATHEANA application.

Although the details of the steps in Figure 2.1-1 are not described until Section 3, the flowchart can be a useful reference as to how the steps relate and where they fit in the overall methodology. Further, although this user's guide discusses the steps in a serial fashion, the steps are typically performed iteratively (as necessary) in any HRA/PRA application. That is, based on the results of a given step, analysts may find that they need to return to some of the steps that they previously performed.

In reviewing the steps, the reader should discover that most of the steps in the ATHEANA methodology are not particularly unique; rather, they represent good practices as part of any HRA. For example, for any technical analysis, the analyst must understand the issue to be addressed and the overall objectives and scope of the desired analysis. Hence, these aspects of ATHEANA should not be viewed as extra work to be performed in an ATHEANA analysis. The ATHEANA methodology simply formalizes the need to perform these steps.

> Many of the steps in ATHEANA are typical good practices and, so, are not particularly unique and do not really represent additional steps in performing an HRA. However, these good practices are formalized as specific steps in the ATHEANA methodology.

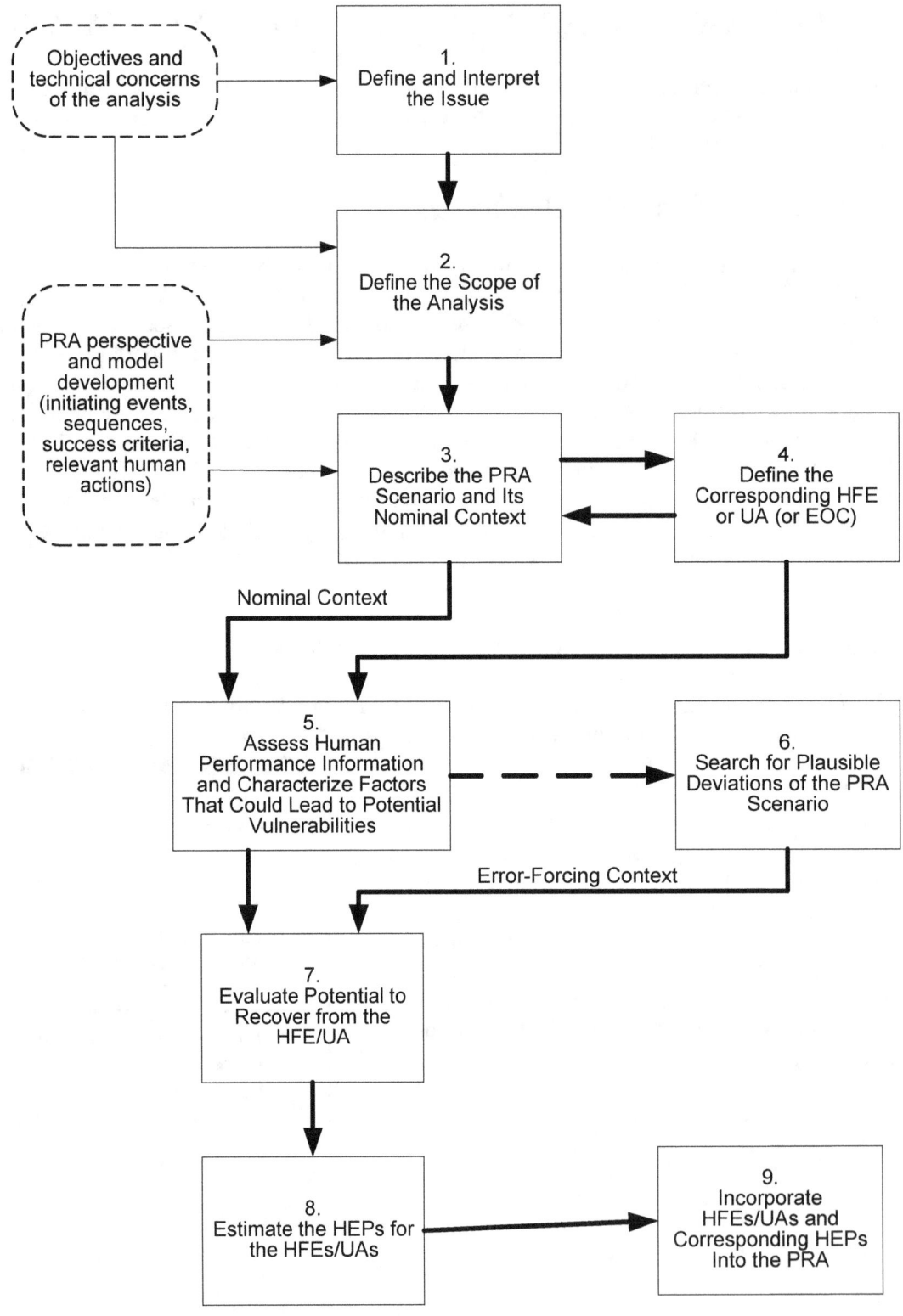

Figure 2.1-1. Steps in the ATHEANA Methodology

Additionally, ATHEANA's steps involving the definition of an HFE and the subsequent determination of the factors likely to have the greatest influence on the probability of operators making the human failure of interest, within the context associated with a particular accident sequence, mirror what is typically done in any HRA. However, in doing so, ATHEANA examines the following considerations:

- Should an HFE be represented by one or more particular UAs, as discussed later?

- Should certain errors of commission (EOCs) also be addressed?

- Should additional aleatory influences, including different plant conditions and other contextual deviations, be considered for the PRA sequence of interest (as discussed later)?

Finally, derivation of the corresponding HEP for an analyzed UA or HFE (i.e., quantification) is performed on the basis of identified important influences on human performance, just as in any HRA quantification technique. The ATHEANA methodology currently uses a formalized expert opinion elicitation process to estimate the HEP, rather than using specific rule sets or similar structures to convert the effects of these important influences into an HEP.

2.2   Scope of Human Events That Can Be Analyzed Using ATHEANA

The current documented ATHEANA methodology is most mature, and its guidance is most specific, for analyzing post-initiating human events in nuclear power plant applications. This user's guide is limited to such applications. However, within this limitation, ATHEANA is meant to be able to address any human action that is potentially important during the response to an initiating event in a nuclear power plant. ATHEANA does consider the need to include EOCs as well as the errors of omission (EOOs) that are typically defined for PRA accident sequences. It also considers the need to represent a PRA-related HFE as one or more UAs (discussed later). Additionally, and partially because of its quantification approach based on expert opinion, ATHEANA can be used to address a wide range of operator performance under all types of conditions such as different plant operating modes (not just full-power), for both pre-core damage as well as post-core damage actions, and for internal as well as external initiating events, as long as the experts have access to experience or a base of knowledge that can be drawn upon to support estimating HEPs for the conditions under consideration.

2.3   Understanding the Key Features of ATHEANA

Section 1 and the above overview of ATHEANA have highlighted the following key features and other defining characteristics of the ATHEANA methodology:

- Consider the need to include relevant EOCs in the PRA, and break down the HFE into specific UAs.

- Identify operational vulnerabilities for the nominal context assumed for the PRA sequence of interest, and for variations in conditions that fit within the PRA sequence of interest and could capitalize on those vulnerabilities and set up potential UAs (i.e., conditions that may make a UA of interest particularly likely or lead to the need to include additional UAs in the model).

- Identify plant conditions, other than the nominal conditions or plant evolution assumed for the PRA sequence of interest (i.e., deviation scenarios), which might cause the above vulnerabilities to make a UA of interest particularly likely.

- Identify important PSFs for both (1) the PRA-defined sequence with its nominal context, and (2) any plausible deviation scenarios.

- Identify other aleatory factors that could significantly affect the likelihood of the human failures and their uncertainties.

2.3.1 *Alternative Representations of the HFE*

In performing an ATHEANA analysis, the analyst considers the need to identify alternative representations of the HFE as originally defined for the PRA, as mentioned in the first distinguishing aspect above. This aspect of ATHEANA, addressed in Step 4 of the process, examines the HFE from two viewpoints — one that considers breaking down the HFE into specific UAs, and another that considers defining one or more EOCs related to the HFE.

Typically, a PRA will model various accident sequences that logically depict how an undesired event such as core damage could occur. For instance, an accident sequence might involve a loss of main feedwater initiating event, followed by a successful reactor scram, but with a failure of all auxiliary feedwater. According to procedure direction and associated operator training, the operators should manually initiate feed-and-bleed cooling upon certain parameter indications (e.g., reactor coolant pressure greater than a certain pressure setpoint when there is insufficient secondary cooling). The operators' failure to initiate feed-and-bleed cooling could be the HFE of interest.

Using ATHEANA, the analyst considers whether there is a need to break down the HFE into specific failure modes called UAs. For instance, for feed-and-bleed, it could be desirable to break down the overall HFE into a "failure to feed" *and* a "failure to bleed" (two related, but different, UAs). Another possible breakdown could be "failure to initiate feed-and-bleed" and "failure to adequately control feed-and-bleed once initiated."

As part of breaking down failure events into specific UAs, it may also be useful to consider different ways in which a given action might be implemented. For example, an analyst may investigate whether there is a risk-related distinction between an operator inappropriately stopping a pump by turning the pump switch to "off," and inappropriately stopping a pump by pulling-to-lock the pump switch. In both cases, the pump is inappropriately stopped (i.e., the same basic human failure); however, in the first case, the pump may be able to subsequently restart upon a renewed actuation signal, whereas in the second case, it will take another operator action to restart the pump (with the operator having to take the pump control out of the pull-to-lock position).

Modeling and analyzing at the UA level provides the means to explicitly investigate the potential impact of different UAs on the plant response, as well as on other human actions. Besides this investigative capability, there could typically be three reasons for wanting to represent the HFE as separate UAs:

(1)   The factors or context that could lead to or otherwise drive the different UAs are identified as being significantly different and it is desirable for the results of the HRA to reflect these differences.

(2)   The various UAs are likely to have very different perceived error rates, including any recovery potential, and it is desirable for the results of the HRA to reflect those differences.

(3)   There is a significant perceived dependency between a particular UA that is associated with the HFE and some other human failure modeled in the PRA (either upstream or downstream

in the chain of events depicted by the PRA sequence). By breaking the HFE into UAs, the specific dependency can be modeled more appropriately and explicitly.

The other part of alternatively representing the HFE involves adding one or more related EOCs (another form of UA). For example, besides having the "failure to initiate feed-and-bleed" in the PRA, the analyst might expand the model to include an EOC such as "operator prematurely stops feed-and-bleed." The HEP associated with this specific act and the factors most important to this specific failure could be different from the originally defined HFE. The reasons for wanting to include an EOC are the same reasons provided above. If the analyst perceives that such an EOC is important to include, that specific EOC should be added to the PRA and analyzed separately with its own HEP.

It should be apparent to the reader that the analyst must anticipate the results of considering the reasons discussed above in order to decide whether an HFE should be represented as specific UAs, including the addition of EOCs. In other words, some level of analysis must be performed. With the limited use of ATHEANA to date, there is inadequate demonstration that breaking down the HFEs into different UAs will be necessary for most applications. However, as will be covered more under Step 4 in Section 3, it is logically reasonable to spend some level of effort examining the potential need to do so. Similarly, while there has not been sufficient use of the method to demonstrate the need to always add an EOC form of the HFE of interest, the need to search for potential EOCs and the contexts that could cause them to occur is becoming more generally accepted [Refs. 4–7] and is certainly recommended.

2.3.2 *Addressing Multiple Contexts for the HFE/UA*

One important commonality among the other distinguishing characteristics of ATHEANA is important to grasp. Specifically, implementing the ATHEANA process involves the following tasks:

(1)   Examine the PRA-defined sequence and the context expected or assumed for that sequence. (In ATHEANA, this is called the "nominal context.")

(2)   Search for and include other plausible conditions that are similar to and fit within the overall PRA-defined sequence, but whose presence make it particularly difficult for the operator(s) to perform the desired action. (In ATHEANA, these are called "EFCs.")

> ATHEANA analyzes an HFE/UA for the context typically associated with the PRA sequence containing the HFE (called the nominal context), *and* for other contexts that may make the HFE particularly likely [called error-forcing contexts (EFCs)].

This difference between the PRA-defined sequence with its associated nominal context, and other ways the PRA-defined sequence may evolve that may induce EFCs, is important to understand and is perhaps best explained by an example.

Depending on the level of detail available from the PRA model, a general understanding of what is going on in the PRA sequence comes from the successes and failures explicitly included in the model itself. However, the HRA analyst needs to understand much more about the accident sequence to perform an HRA for the HFE (or UA in ATHEANA) of interest. For instance, the HRA analyst needs to fully understand the procedure directions and the extent to which the operators are trained on that sequence. The HRA analyst also needs to understand how the plant responds thermal-hydraulically in such a sequence, and what parameter indications or other cues will occur and when, in order to estimate such things as diagnosis and implementation time to perform the action of interest.

Additionally, the HRA analyst (typically with PRA modeling input) makes certain assumptions to complete the understanding of the accident sequence. For instance, equipment failures are typically assumed to occur in a complete fashion (i.e., the loss of main feedwater is assumed to be complete and all at once, rather than being lost in a gradually degrading fashion, and the failure of auxiliary feedwater is assumed to be "all at once," rather than operating in a partial or degraded state for a time before failing). Further, the HRA analyst often assumes that all operating crews have a level of homogeneity that allows the crews to be treated the same from a performance point of view. Also, the HRA analyst typically assumes that all instrumentation is available and working properly during the accident sequence unless there is a specific reason to assume otherwise. Such assumptions are consistent with typical PRA modeling interpretations and practices.

The above examples illustrate what the HRA analyst needs to understand before he/she can determine what factors may be most important to the HFE/UA of interest (in this case, operators failing to initiate feed-and-bleed), and what the probability of failure is estimated to be. In ATHEANA, putting all this information together is what is meant by "describing the PRA scenario" (Step 3 of the methodology) and its associated nominal context with the HFE or UA in mind (Step 4). After defining and understanding all of this context, the HRA analyst determines what factors may introduce vulnerabilities for the HFE/UA of interest, given this nominal context (Step 5). For example, the training for this type of sequence may be infrequent or inadequate in some way, or it may be difficult to read or interpret a key indicator that is needed to know when to take the desired action for the PRA scenario. In this step, the analyst also qualitatively evaluates the effects of these factors on the performance of the desired action, given the nominal context [e.g., the time available for the action is short and, thus, this factor will have an appreciable impact on the likelihood of the HFE/UA (i.e., the HEP)].

To complete the evaluation for the nominal context associated with the PRA sequence, Step 6 is momentarily skipped and, in Step 7, the analyst considers the reasons for and the likelihood of the operators being able to recover quickly from the HFE/UA before undesired consequences occur. With the knowledge from these previous steps, an HEP, including consideration of recovery, is estimated in Step 8 (in this case, for the failure to initiate feed-and-bleed).

Using ATHEANA, the analysis is performed for this model of the basic PRA sequence and its associated nominal context as described above. But then, in ATHEANA, there is an additional pass through some of the steps of the methodology in search for plausible conditions (i.e., EFCs) that may make the HFE/UA particularly likely.

In this second pass, a portion of the analysis is repeated. Taking advantage of the knowledge about the vulnerabilities previously defined in Step 5, Step 6 is performed (shown as the dashed line from Step 5 to Step 6 in Figure 2.1-1) and the analyst proposes alternative ways that the PRA sequence could proceed. That is, in Step 6, the analyst investigates ways that plant conditions may evolve (other than that assumed for the nominal context), including other aleatory and potentially complicating factors. For instance, in this case, using ATHEANA guidance for Step 6, the HRA analyst is encouraged to investigate the potential effects of differences in context such as the following:

- Is it plausible, and would it matter to the HFE/UA, if the sequence evolved somewhat differently, such as auxiliary feedwater first operating in a partial/degraded state before failing, thus allowing the possibility that the crew will pass by the feed-and-bleed step in the procedure, but then have to come back to that step once auxiliary feedwater is completely lost?

- Is it plausible, and would it matter to the HFE/UA, if the sequence evolved so that by the time it was necessary to initiate feed-and-bleed, the auxiliary feedwater equipment failures were found to be minor and expected to be easily recoverable, thus allowing the possibility that the operators would fail to begin feed-and-bleed when they should (i.e., in anticipation of restoring auxiliary feedwater momentarily)?

- Would it matter to the HFE/UA if there are known differences in crew characteristics at the plant? For example, are there differences in crews such that some operate in a very slow and methodical manner, while others are typically much more rapid and anticipatory when implementing a procedure, and it appears that for the slow crews the time to diagnose and/or implement the desired action may not always be sufficient?

- Is it plausible, and would it matter to the HFE/UA, if a key indication used to determine when to initiate feed-and-bleed happened to be unavailable (such as for calibration, or as a result of a common workaround) or it failed shortly after the scenario began?

- Is it plausible, and would it matter to the HFE/UA, if a number of other miscellaneous failures and associated alarms occurred during the sequence (e.g., an air compressor happened to fail or the recirculation alignment of the condensate pumps did not automatically occur correctly) because these might create diversions of operator attention and/or slowdown the operators response to the event?

What is being done by these investigations is to determine if there are one or more plausible deviations of the PRA accident sequence, such that the operators may find it particularly difficult to perform the desired human action within the necessary time. Again, such deviations are said, in ATHEANA, to have EFCs. The search for such deviations from the PRA sequence is performed, as the reader may recall from Section 1, because experience strongly indicates that well-trained operators usually make serious errors only when the situation is unusual or difficult to understand, or otherwise sets up the operators for failure. This formal consideration of deviation scenarios, including changes in plant conditions and other aleatory influences (e.g., crew differences) and their possible effects on the HFE/UA, is at the heart of ATHEANA.

If plausible (i.e., not too unlikely) and seemingly troublesome contexts are identified, the methodology then calls for the analyst to take the results of Step 6 and retrace through Steps 7 and 8 (shown as the dashed line feeding from Step 6 into Step 7 in Figure 2.1-1) and perform those steps again, but this time with consideration of what appears to be plausible EFCs from Step 6. If these other contexts are indeed EFCs, the corresponding estimated HEP(s) will likely be higher than the HEP for the nominal context.

2.3.2 The Mathematical Treatment of Multiple UAs and Contexts, Including Incorporation Into the PRA

Because of the potential breakdown of an HFE identified in a PRA (e.g., failure to initiate feed-and-bleed) into specific UAs (e.g., failure to initiate, initiate but prematurely shutoff), each HFE may be represented by one or more UAs, and each UA could occur under one or more contexts. These contexts could include the nominal context, as well as other EFCs subsequently developed in Step 6. In the simplest case of one UA for the HFE modeled in a PRA accident scenario (S), the UA could still be analyzed for several contexts (nominal and one or more EFCs), so quantification of the HFE is calculated as follows:

$$P(HFE|S) = \sum_i P(EFC_i|S) \times P(UA|EFC_i,S)$$

In other words, the probability of the error for the HFE applicable to accident scenario "S" is equal to the summation, across all contexts (each context is shown as EFC_i in the equation, but this is intended to also include the nominal context) associated with scenario "S," of the products of the probability of each EFC_i for scenario "S," times the conditional probability of the UA error rate (i.e., the HEP for the UA) given each corresponding EFC_i.

For the even more general case of multiple UAs applicable to several contexts, the above equation can be written as follows:

$$P(HFE|S) = \sum_j \sum_{i(j)} P(EFC_i|S) \times P(UA_j|EFC_i,S)$$

Here, we have the additional summation over multiple UAs, as appropriate, for any given context EFC_i.

To illustrate the above equations, suppose a PRA sequence has two very different associated contexts (based on carrying out the ATHEANA steps), such that the sequence is expected to evolve as nominal context #1, 90% of the time and, for the remaining 10% of the time, the sequence is expected to evolve according to a much more challenging deviation scenario (context #2). Further, assume that two UAs are being evaluated for each context; one is the EOO of "failure to initiate feed-and-bleed" (UA #1), and the other is the EOC of "premature shutdown of feed-and-bleed" (UA #2). [Note that a context that might induce an EOC would typically be different than a context that might induce an EOO; however, we have intentionally kept this example simple for illustration purposes]. Based on subsequent determination of the HEPs by the expert elicitation process for the UAs being addressed, considering the plant conditions and relevant PSFs associated with each context, suppose the following results are obtained for the HEPs:

- The HEP for UA #1 given context #1 is estimated as 1E-2.

- The HEP for UA #1 given context #2 is estimated as 1E-1.

- The HEP for UA #2 given context #1 is estimated as 1E-4.

- The HEP for UA #2 given context #2 is estimated as 1E-2.

Then, according to the equation, the overall HEP for the HFE for the PRA is equal to [(0.9 x 1E-2) + (0.9 x 1E-4) + (0.1 x 1E-1) + (0.1 x 1E-2)] or 2.009E-2 or ~2E-2.

As can be seen, a rigorous quantification of the overall PRA-defined HFE is a two-step process:

(1)   Estimate the likelihood of each context.

(2)   Estimate the HEP associated with each UA for each context for which the UA applies.

The probability of each context can generally be quantified using typical PRA tools of systems analysis and data. For instance, suppose the difference between contexts #1 and #2 (above) is that context #2 involves that portion of the plant trips whereby there are numerous extraneous or less-important equipment failures and associated alarms that can slow down operator performance and/or present significant diversions of resources during the response to the scenario. The probability that this occurs 10% of the time might be based on experience with plant trips and/or typical equipment failure data and modeling (correspondingly yielding 0.9 as the probability that the nominal context exists with no significant extraneous failures/alarms). Sometimes, estimating the probability of a context may require expert elicitation techniques similar to that described for Step 8 in Section 3.

Finally, in Step 9, the results are all incorporated into the PRA. This may be achieved in two ways. The first way involves maintaining the PRA logic model and original defined HFE. The frequency of the accident sequence excluding the HFE is determined the typical way, based on the probabilities of the successes and failures associated with the sequence including the frequency of the initiating event. The HEP for the HFE would be as determined above; in this case, the value would be ~2E-2. The second way is to expand the original PRA modeled sequence to explicitly reflect the different contexts and specific UAs (including any EOCs) for each context. This might be done, for instance, in either the event trees (by adding top events) or the fault trees (by adding basic events). The HEPs would then be applied to the UAs as appropriate. Figure 2.3-1 provides an illustration of both ways to incorporate the results into the PRA.

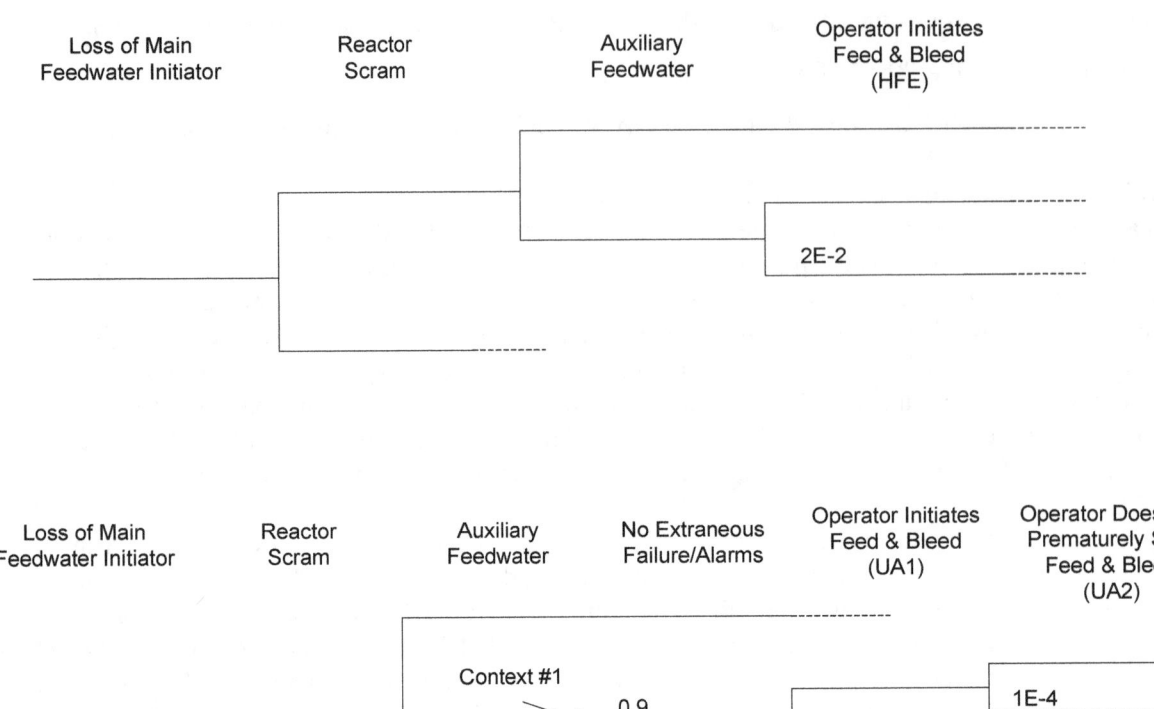

Figure 2.3-1. Two Ways To Incorporate ATHEANA Results into the PRA Model

2.4   Advantages of Using ATHEANA

The use of ATHEANA is meant to provide the following advantages:

(1)   a focused prediction of the specific error that might be made and the most influential factors affecting that specific error

(2)   increased assurance that the major risk associated with the HFE has indeed been captured

(3)   the ability to estimate HEPs for all sorts of combinations of factors and various conditions

In general, the advantages are achieved because of the following aspects of the method:

(1)   the possible redefining of the HFE into specific acts of concern (i.e., UAs), including possible EOCs

(2)   the method's forced in-depth understanding of context associated with an HFE/UA, including consideration of many factors that may affect human performance

(3)   the search for and analysis of plausible EFCs and the resulting effects on the HEP

(4)   the use of a formal expert elicitation approach to estimate the HEPs[5]

In the case of the first advantage regarding the breadth and depth of the analysis, for instance, an ATHEANA analysis might find that "failure to initiate feed-and-bleed" (the general description of the HFE, according to the PRA model) is, in fact, more likely to occur because the operators (1) fail to initiate it in time; (2) fail to properly control it, thus starving the core of sufficient flow; or (3) prematurely shutoff feed-and-bleed. Further, by examining a wide range of factors that might affect human performance, clearer understanding should be possible as to what form of the failure is most likely to occur, and the most prominent influences affecting the predicted failures. This deeper understanding should allow a more focused determination of the specific lessons learned, if any, that could be implemented to improve the predicted operator performance.

In the case of the second advantage regarding assessment of risk, understanding how the frequency of a context and the associated HEP go together (as demonstrated in Section 2.3.3) can help to increase assurance that the major risk associated with the HFE has indeed been captured. For instance, for the typically applied nominal context associated with a PRA sequence, it is usually assumed that there are no extraneous failures or other aleatory influences beyond that already explicitly specified or directly inferred from the successes and failures in the PRA model. Thus, the frequency of the context is equivalent to the frequency of the PRA sequence leading to the HFE of interest. So for example, if the PRA sequence involves loss of main feedwater as the initiating event at an estimated 0.1 per year, and failure of auxiliary feedwater estimated at 1 E-4 per demand, with all other success probabilities nearly 1.0, the frequency of the sequence/nominal context up to this point is 1 E-5 per year. If, for this nominal context the HEP for the HFE of interest is estimated to be 5 E-3, the overall combined frequency for the sequence including the HFE becomes 1 E-5 per year x 5 E-3, or 5 E-8 per year.

Now, let's suppose that ATHEANA's search for EFCs reveals that if the PRA sequence were to occur with other less-important alarms and extraneous failures beyond those associated solely with the system failures explicitly identified by the PRA sequence, and if a very methodical crew (rather than a faster-acting anticipatory crew) were to be on-shift at the time, the result could be a much slower crew response. This might be caused by the crew spending time and diverting resources to determine the nature and severity of the extraneous alarms/failures, as well as the slower response of the most methodical crews at the plant.

---

[5]   This aspect allows for a quantification process that is flexible and not tied to specific rule sets. Such a process can cover various combinations of influencing factors and possible dependencies among the factors, as long as the experts have access to experience or a base of knowledge that can be drawn upon to support estimating HEPs for the conditions under consideration. However, this flexibility does result in the need for disciplined and thorough documentation of the results of the quantification process, so that the basis and derivation of the HEPs is clear.

For illustrative purposes, suppose it is determined, on the basis of past real challenges, that extraneous alarms and failures sufficient to slow crew response occur about 50% of the time. Further, suppose that 25% of the crews at this plant are particularly methodical in their response tactics, demonstrating a measurably slower response than the rest of the crews. Considering the possible delays and attention diversions caused by these two effects taken together, for this illustration, assume that the resulting HEP is estimated to be quite high, say 0.3 (i.e., there is an estimated 30% chance that the crew would not reach or otherwise implement the feed-and-bleed step in the procedure to prevent core damage).

For this context involving the PRA sequence, but with additional extraneous alarms/failures and with a methodical crew, the frequency of the context becomes 0.1 per year (loss of main feedwater initiator) x 1 E-4 (failure of auxiliary feedwater) x 0.5 (probability of extraneous alarms/failures) x 0.25 (probability that a very methodical crew is on-shift at the time of the event) = 1.25 E-6 per year. Combining this frequency with the HEP for this set of conditions yields 1.25 E-6 per year x 0.3 = 3.75 E-7 per year. This result is about 7 times greater than that determined by looking solely at the nominal context (above), and represents the risk that was overlooked by considering only the nominal context.

Of course, whether this advantage is actually realized depends on whether EFCs that are not too unlikely can be typically identified. Nonetheless, the corresponding HEP becomes sufficiently high as to more than offset any change in the context frequency going from a nominal context to a more specifically defined and particularly challenging context. However, without investigating the potential for — and effects of such EFCs, it cannot be known if analyzing only the typical nominal context for a PRA sequence is indeed capturing most of the risk associated with a particular HFE. In ATHEANA, this investigative process is explicitly included in performing an HRA.

As for the final advantage, the ability to estimate HEPs for combinations of influencing factors for various contexts, use of an expert elicitation approach provides a flexibility not typical of quantification methods that use specific rule sets or other prescribed structures for converting the qualitative judgments about the effects of these influencing factors into HEPs. Typically, and to keep the analysis simple, many quantification techniques treat the influencing factors as having independent effects on the HEP, and they use specific multipliers, curves, or other pre-established means for estimating the HEP. By contrast, the ATHEANA expert elicitation process allows dependency effects among the influencing factors to be considered, to the extent considered important by the experts, and the estimation of the HEP is not tied to the same multiplier or quantitative effect for a factor regardless of changes in context. However, as previously noted, this approach requires disciplined and thorough documentation of results.

2.5   When To Use ATHEANA

2.5.1   *General*

Considering the advantages versus the corresponding effort required to perform an ATHEANA HRA, a logical question is, "when should I use ATHEANA?"

Because there have been only a limited number of ATHEANA applications to date, it is not possible to definitively answer the above question on the basis of numerous trial applications. Nonetheless, it should be apparent that ATHEANA investigates human failure potential for a range of contexts beyond the nominal context including identifying EFCs, considers a broad number of PSFs (not just a set few PSFs) for all analyzed contexts, and distinguishes among different but related human failure modes. Hence, based on these features, it can be suggested that it is most appropriate to use ATHEANA when a risk-informed decision requires one or more of the following characteristics:

- It is important to analyze and understand vulnerabilities associated with specific human failure modes (i.e., UAs) including possible EOCs, rather than using a simple and more general description of the HFE and one that may consider only an EOO.

- It is important to perform a thorough investigation of different contexts, including different plant conditions and plausible aleatory influences that may affect the HFE/HEP evaluation so that the decision-maker is informed about possible EFCs (although they may be less likely than the nominal context) for which the HRA results could be quite different.

- It is important that the HRA consider a wide range of potential influencing factors for a number of different contexts to ensure that the most important influencing factors are identified and explicitly considered in the HEP evaluation.

> Consider using ATHEANA when the analysis objectives require that the results include any of the following considerations:
> ✓ the important specific human failure mode(s)
> ✓ plausible contexts that may be particularly troublesome or difficult, and the resulting HEPs
> ✓ a wide range of PSFs across multiple scenario contexts

The above characteristics could be particularly important or relevant when one of the following conditions exists:

- A relatively complex design or procedural change is part of a risk-informed submittal, especially if the change could increase the potential for an EOC or have a strong impact on human performance.

- An existing PRA/HRA is being updated, and resources are available for a more detailed analysis.

- An initial or screening HRA analysis has been performed and importance measures or other evaluations have identified HFEs that are particularly important to preventing core damage, or sensitive to changes in HEP values (see the discussion in the next section).

2.5.2   Screening for When To Use ATHEANA

While an analyst could certainly choose to apply ATHEANA for any or all HFEs associated with a given analysis up to and including, for instance, all HFEs in a full-scope PRA, the additional effort may not always be warranted. Thus, to further determine when it may be most appropriate to apply ATHEANA, the following are suggestions for deciding when to apply the ATHEANA methodology to particular HFEs. These suggestions assume that there is already an existing HRA done with (1) other HRA methods (which is likely to be the case for many current applications of PRA/HRA), or (2) the use of approximate or reasonable screening values to obtain preliminary risk results. These suggestions should help the analyst decide when application of ATHEANA might prove useful and potentially make the additional effort worthwhile:

(1)   If the HFE evaluation already yields a high HEP (e.g., >0.3), it is less important to search for additional contexts that may make the HEP even higher. Expending resources for an ATHEANA analysis for such HFEs could be of marginal additional benefit.

(2)   Review the importance measure results for the HFEs in question, especially the risk achievement worth or risk achievement interval values, to determine the HFEs for which the quantitative analysis results are most sensitive. It is more desirable to apply ATHEANA to the most significant of these HFEs, particularly to determine whether the interpretation of the results might be affected by considering specific UAs, including EOCs, and additional EFCs.

(3)   Perform sensitivity evaluations of the HFEs, singularly and in logical groupings (because the HFEs are related), to determine how much change in the HEPs would be required to change the results in a significant way (e.g., quantitative results; the risk ranking of important sequences, components, or HFEs). ATHEANA may be appropriate for those HFEs that could produce significant changes in the results for plausible changes in the HEPs.

(4)   Determine those HFEs and their uncertainty estimates that have the most impact on the overall uncertainty of the results (and the decisions to be made considering that uncertainty). ATHEANA may be appropriate to apply to these HFEs because of the additional uncertainties that may be manifested, especially when additional context considerations are included.

By using such screening techniques, ATHEANA can then be applied to those HFEs that would most benefit from using the ATHEANA methodology.

As to which HFEs may benefit from application of ATHEANA, consider the following:
✓ whether the HEP could realistically go much higher
✓ the importance rankings of the HFEs
✓ the results of HEP sensitivity evaluations
✓ possible effects on uncertainty of the overall result

3. STEP-BY-STEP GUIDE FOR PERFORMING THE ATHEANA PROSPECTIVE ANALYSIS

Before beginning the specific steps of the prospective analysis, there are two general preparatory tasks that need to be performed. These preparatory tasks are identical to those that would be performed in preparation for any HRA method, but there are a few aspects of the preparations that are unique to ATHEANA and warrant special notice.

Assembling and Training the Analysis Team

ATHEANA should be applied by a multi-disciplinary team under the leadership of the HRA analyst. However, a full team of experts, as described below, will not be needed at all times during the application of an HRA in the context of a PRA. The HRA analyst, in conjunction with the lead PRA analyst, can perform much of the analysis on their own, collecting relevant information from others as needed. An important point to obtain from the following discussion is that at critical times, certain types of expertise will be essential to complete an adequate analysis. Thus, the HRA analyst, in conjunction with the PRA team lead, will be required to organize the HRA and schedule discussions with the appropriate individuals, in order to ensure that all of the relevant information is obtained. It is essential that the HRA and PRA analysts have access to people with sufficient knowledge and experience to supply the information and answer the questions involved in the ATHEANA process.

The HRA analyst must have access to the following expertise to most fully implement ATHEANA:

- understanding of the ATHEANA underlying basis and process steps sufficient to grasp its principles and implement its searches

- knowledge of the plant-specific PRA, including knowledge of the event sequence model to be familiar with what is modeled as having been successful, as well as what has failed, in the scenario(s) of interest

- understanding of plant behavior, especially thermal-hydraulic performance, because this provides knowledge about the timing of events and indications (or lack thereof) of cues for operator actions

- understanding of the plant's procedures and operational practices because these provide the formal and informal guidance used by the operators

- understanding of operator training and training programs because these provide the underlying expectations of how and when operators perform their duties

- knowledge of the plant's operating experience, including trip and incident history, backlog of corrective maintenance work orders, etc., because this can provide information about recent problems, difficulties, experiences, and biases that may affect future operator performance

- knowledge of plant design, including man/machine interface issues inside and outside the control room because both problematic and ideal conditions in this area can affect the likelihood of operator success

- knowledge of the importance of controlling for bias in performing expert elicitations, to support the facilitator-led, expert opinion elicitation process used for quantifying the HFEs.

Therefore, it is recommended that the general analysis team include the following technical staff:[6]

- an HRA analyst

- a PRA analyst (preferably the accident sequence task leader)

- a reactor operations trainer (with expertise in simulator training)

- a senior reactor operator

- a thermal-hydraulics specialist (usually supporting the PRA as needed)

- plant personnel responsible for emergency procedures (when possible)

Other plant experts should supplement the expertise of the analysts as needed, to provide additional plant information required for the ATHEANA process, participate in simulator trials or talk-throughs, and support the collection of information needed to quantify the HFEs.

The HRA analyst serves as the team leader, and also is the principal expert on human performance issues in the behavioral sciences, the ATHEANA knowledge base, and the ATHEANA process. In particular, the HRA analyst must perform the following functions:

- Provide interpretation and guidance to the team as needed, in order to ensure that the analysis meets the the objectives of ATHEANA, and of the HRA and PRA overall.

- Facilitate the collection of information needed to supplement the experience and expertise of the team.

- Collect or facilitate the collection of information needed to quantify the HFEs identified.

- In some cases, serve as the facilitator for the expert judgment quantification process (note that in some cases, it may be more appropriate for the PRA analyst, with significant plant knowledge and understanding of plant behavior to be trained to perform this function)

- Familiarize other team members on ATHEANA, its underlying principles, and other relevant HRA areas sufficient to be meaningful participants (e.g., controlling bias during expert elicitation).

Collecting Background Information

Success in applying any HRA method depends upon the quality of information collected and used in the analysis. Consequently, information collection is an important HRA activity and is typically performed throughout the analysis.

Application of the ATHEANA HRA method requires the same kinds of information (e.g., plant-specific design, procedures, operations, and maintenance related information) that are collected in other HRA methods. In addition, as the analysis proceeds, the ATHEANA HRA analyst should collect plant-specific and/or issue-specific information (especially operational experience) that is needed to meet the specifics of the method (e.g., identifying informal rules on which operating crews might rely and important decision points in procedures).

---

[6] This is with the understanding that the HRA and PRA analyst can perform much of the work, obtaining information from individuals as needed, and getting the whole team together only at critical points during the analysis.

The types of background information necessary to apply ATHEANA are described in the appropriate steps below. For a general summary of this information, see Section 7.3 in NUREG-1624 [Ref. 1].

In addition, although the critical relevant knowledge-base is provided in this document, it will likely be beneficial if the HRA analyst is familiar with the ATHEANA knowledge-base presented in NUREG-1624, particularly the ATHEANA multi-disciplinary HRA framework (Sections 2 and 3), the behavioral science perspective on human error (Section 4), and the analyses of operational events from the ATHEANA perspective (Section 5).

3.1   Step 1: Define and Interpret the Issue

3.1.1   *Purpose*

The purpose of this step is to give the analysts a clear understanding of the issue before them. In other words, in this step, the analysts define the objective that is to be achieved by performing the HRA.

3.1.2   *Guidance*

The analysts perform three tasks to define and interpret the issue:

(1)   Identify the source of the issue.

(2)   Clearly define the issue in technical terms relevant to HRA.

(3)   Further interpret the issue in terms useful to perform the HRA/PRA (i.e., in a risk framework).

3.1.2.1   Task 1.1: Identify the Source of the Issue

The ATHEANA analysis begins when the analysts are tasked to address an issue related to the impact of human performance on risk. Such an analysis request could come from the following sources (among others):

- regulators, government officials, or other authoritative persons
- management personnel
- other technical staff
- members of the public

Knowing the source establishes the intended audience to whom the HRA results (an issue resolution) will ultimately be provided. This, in turn, can help in defining how (and at what level of detail) the analysis results should best be presented [e.g., summarized in a non-technical manner, with (or without) technical jargon and other terminology unique to the activities related to the issue].

3.1.2.2   Task 1.2: Define the Issue

Issues are often initially expressed in somewhat vague terms, lacking a technical perspective most relevant for the HRA. For example, the issue might be originally expressed as "How is safety affected if training on emergency operating procedure X is changed to once every 3 years, as opposed to the current frequency of once per year?" For purposes of the HRA, the issue should be redefined in terms of a technical analysis that is germane to what HRA can potentially assess. For instance, in this case, a more technical statement of the issue might be, "What effect will there be on operating crew readiness and performance, as measured by the expected change in crew reliability, when faced with having to implement emergency procedure X, if the frequency of training on emergency procedure X is changed from once per year to once every 3 years?"

The goal of this redefinition is for the source and the analysts to agree on a clear and preferably concise statement of the issue in unambiguous terms amenable to the HRA. This statement of the issue might be as short as a single sentence, or might require much more verbiage. However, brevity is encouraged to the extent possible. Agreement on this definition is to put the issue source and the analysts "on the same page" to minimize misinterpretation regarding the issue that the analysts plan to address when performing the HRA.

3.1.2.3   Task 1.3: Further Interpret the Issue in a Risk Framework

With the source of the issue and an HRA-relevant technical statement of the issue in hand, further interpretation of the issue is likely to be needed. This task within Step 1 is for the benefit of the analysts themselves, al though it may also be useful to the source of the issue. This additional interpretation involves understanding the issue in terms of how the analysts will have to use the HRA/PRA (the actual scope of the analysis is covered in Step 2) to address the impact of the issue in terms of risk. For instance, the analysts are likely to need to understand the answers to questions like the following:

- What risk metric is the analysis concerned with (e.g., core damage frequency, large early release frequency, average yearly risk, change in instantaneous risk)?

- Is the issue only affecting the likelihood portion of risk, the consequence portion of risk, or both?

- Is this an issue that will affect many different operator actions under many different conditions, or does it, for instance, affect only one human action for one very specific set of conditions?

- Is the issue germane only to power operation, or are low-power and shutdown modes also relevant?

> In this step, the analysts are to perform the following tasks:
> - Identify the audience to whom the issue resolution is to be provided.
> - Define the issue in HRA terms.
> - Provide an overall risk framework for resolving the issue.

Having this risk-relevant view of the issue provides a risk framework, involving HRA/PRA, for addressing the issue. It also provides an initial step in defining the scope of the analysis that is performed in Step 2.

3.1.3 *Output*

Step 1 yields the following output:

(1)   identification of source of the analysis request (i.e., the audience for the issue resolution) and, hence, how the results should best be presented

(2)   a statement of the issue to be addressed, agreed upon by the source of the issue and the analysts, in technical terms relevant to HRA/PRA

(3)   an understanding of what risk impacts are relevant to the issue being addressed

3.2   Step 2: Define the Scope of the Analysis

3.2.1 *Purpose*

The purpose of this step is for the analysts to decide on the scope of the HRA that will be performed to address the issue defined in Step 1. This sets the boundaries of the analysis so that the analysts understand what is, or is not expected to be included in the HRA.

3.2.2 *Guidance*

The analysts perform two tasks to define the scope of the analysis. This guidance assumes that there is either (1) an existing overall PRA model or other risk-related framework within which the HRA results will be a part, or (2) a PRA model or other risk-related framework will be constructed, if one does not already exist, to provide the risk-relevant impact of the HRA results. These two tasks are as follows:

(1)   Define the scope of the analysis considering a number of analysis scope elements so as to be able to provide the impact of the issue in terms of risk.

(2)   Further limit the scope of the analysis, if possible, by prioritizing what portions of the analysis are required to obtain HRA results sufficient to address the issue.

3.2.2.1 Task 2.1: Consider a Number of Scope Elements

In this task, a first cut at defining the scope of the analysis is performed as measured by a number of elements amenable to a PRA or similar framework. These elements are posed here in the form of illustrative questions. They are meant to serve as aids to the analysts in determining the scope of the analysis, and can be modified as needed to fit the nature of the issue. For instance, the following questions are illustrative of those most relevant to an operating nuclear power plant risk analysis (but they would likely have to be modified to fit, for example, a spent fuel pool concern, an aircraft pilot activity, or a medical procedure type error):

•   What plant operating modes have to be included in the analysis?

•   What plant safety functions have to be addressed in the analysis?

•   What initiating events (both internal and external) have to be included in the analysis?

•   What systems and/or equipment have to be included in the analysis?

- What other human actions (besides those directly associated with the issue) have to be included in the analysis?

- What scenarios and/or sequences have to be included in the analysis?

- What failure modes (of equipment and human responses) have to be included in the analysis?

- What levels of modeling detail have to be included in the analysis?

- What, if any, other characteristics are needed to define the scope of the analysis?

These questions need to be answered with careful consideration of the issue, the human action(s) involved, and the risk-related impacts to be addressed. For example, if the issue and human actions involved are related to activities that would be performed by nuclear power plant operators only post-core damage in an accident sequence (i.e., the issue impacts the Level 2 portion of the PRA, but not the Level 1 PRA), the scope of the HRA may be such that the analysts can focus solely on those post-core damage functions of interest, while using certain Level 1 PRA results simply as inputs into the HRA.

The answers to some questions may already be self-limiting based on the definition of the issue. For example, if the issue is solely associated with determining the impact of a human action on the frequencies of certain specified accident sequences [e.g., only small loss-of-coolant accidents (LOCAs)] during only shutdown conditions (not at power), then other sequences and plant operating modes need not be included, by definition. Similarly, the issue could limit the analysis to specific failure modes of certain specified equipment, such as assessing the impact of a human action only when a particular high-pressure pump fails as a result of a loss of room cooling.

Levels of detail should also be considered. For instance, the issue and human action of interest may be such that certain pipe failures and their probabilities of failure have to be included in the analysis, although pipe failures are normally not included at all in the typical risk framework.

On the other hand, the issue could imply a very broad scope, involving many different human actions, across a wide variety of sequences, involving the total core damage frequency. Further, the analysis may require that additions be made to an existing model or framework (e.g., new types of sequences or initiators) in order to be able to determine the risk impact.

Any assumptions that affect the scope of the analysis should also be evident in establishing the scope of the HRA to be performed.

3.2.2.2 Task 2.2: Prioritize What to Include in the Scope of the Analysis

To further ensure both manageability and efficiency, it is important to perform the analysis using only those resources needed to sufficiently address the issue. Consequently, it is recommended that analysts use this step to prioritize the results of the previous task.

This prioritization could be performed either qualitatively or quantitatively. In the end, it involves anticipating what will be important to include within the scope of the analysis to sufficiently address the issue. For example, it may be that the human action of interest is relevant to many different scenarios/sequences involving LOCAs of various sizes.

However, it may be apparent that the overwhelming majority of the risk impact of the HRA can be sufficiently captured to meet the needs of the issue by examining only the small LOCA sequences because of their frequencies relative to the frequencies of other LOCA sequences. Hence, the analysts could choose to assign the small LOCA impacts as the highest priority of the HRA and even limit the analysis solely to those types of sequences.

> In this step, the analysts are to perform the following tasks:
>
> - Consider a number of analysis elements in defining the scope of the analysis.
> - Prioritize what is necessary to do to limit the analysis while efficiently meeting the needs of the issue.

This prioritization, if not obvious, can be aided by performing sensitivity studies to examine the potential impacts of a range of possible HRA results. Doing so should ensure that the sensitivities are sufficiently robust so that the resulting scope of the analysis will adequately determine the quantitative impact of the issue without missing important qualitative insights that could, for instance, apply to a limited portion of the overall risk.

3.2.3 *Output*

The output of Step 2 consists of a definition of the scope sufficient to understand the boundaries of what the analysis will include. Assumptions made in defining the analysis scope should also be apparent as part of the scope definition.

3.3   Step 3: Describe the PRA Accident Scenario and Its Nominal Context

3.3.1 *Purpose*

The purpose of this step is for the analysts to describe each accident scenario (referred to as "the scenario" from here on) relevant to the issue, as well as the human action(s) to be addressed by the HRA. This description provides the analysts with a working understanding of each scenario, and its nominal context, that is associated with the human action(s). This description is provided at a level consistent with what is typically represented in most PRAs, but with additional detail pertinent to performing an HRA. The analysts will add even more details about the context, in Step 5, when the context is examined in light of factors that may affect performance of the human action(s).

While the following guidance is written for a single PRA scenario, a human action of interest to the issue may be included in multiple scenarios. For example, the identical action involving failure to control safety injection may apply to various transient-initiated scenarios, as well as small LOCA scenarios. While similar, these various scenarios may involve slightly different timing of cues and other aspects that could affect the control of safety injection. In such cases, this step can be performed for one scenario at a time, or for groups of scenarios with similar characteristics.

To the extent possible, analysts should provide scenario descriptions for groups of scenarios, while noting slight differences among the scenarios in a group. This approach is recommended because the same applicable descriptive information need only be developed once, providing efficiency in the process. However, to the extent that the resulting HEPs (and what drives them) might be significantly different, the context of each scenario may need to be kept separate from the others until the analysts decide (during quantification) whether a given HEP (and its drivers) can be applied to groups of scenarios.

3.3.2   *Guidance*

The analysts perform the following tasks to describe the scenario and its context:

(1a)   Develop a description of the scenario in terms of what is readily discernible from the PRA or other risk-related framework that provides an initial model of the sequence of events involving the human action(s) of interest.

(1b)   Augment the above information by collecting, assimilating, and including (in the scenario description) additional information about the scenario, as needed, so that its nominal context can be understood and examined in Step 5 in light of the factors that could affect performance of the human action(s).

(2)   Document this description for use in subsequent steps and especially when the expert elicitation process is followed in Step 8 to quantify the HEPs. In that step, the experts need to understand the contexts for which the HEPs are being assessed, and having this documentation will greatly facilitate that process.

The reader should note that the analysts may perform Step 3 in an iterative manner, or even concurrently with Step 4 regarding definition of the HFE. For instance, this may be particularly true if the HFE of interest is already well-defined by the issue, and the analysts are determining what types of scenarios are most relevant to the human action(s) of concern. Nonetheless, experience with ATHEANA suggests that the analysts will likely find it useful to perform Steps 3 and 4 in parallel so that the scenario description and the human action(s) of interest are considered together.

Note that Tasks 1a and 1b (above) are described below as a single activity (Task 3.1) because it is difficult to separate whether the descriptive information is actually coming from the PRA or from some other source. What is important is that the description address the characteristics discussed below, as these will likely be relevant to performing the HRA.

3.3.2.1   Task 3.1: Develop a Description of the Scenario Based on the PRA (or Similar Framework) and Other Sources

When developing the scenario description, the HRA analyst(s) will likely need to interact with at least the PRA analyst/modeler. The overall goal is to develop a description of the scenario that is (1) consistent with the modeling of the scenario in a PRA or similar framework, and (2) useful for the HRA. If an existing PRA or other model of the scenario is not available or an existing one requires considerable modification to meet the needs of the issue, the HRA analyst should support the PRA analyst to the extent appropriate in developing the scenario model structure. The analysts should then develop a scenario description from the information developed through that exercise.

3-8

By "scenario description," we mean that which is likely to be available, for example, based on an analyst documenting an accident sequence in response to the AS-C2 requirement in the ASME PRA Standard [Ref. 2], but with more detail pertinent to the HRA. From Section 2, the reader will recall the statement that depending on the level of detail available from the PRA model, a general understanding of what is going on in the scenario comes from the successes and failures explicitly included in the model itself. However, the HRA analyst needs to understand much more about the accident sequence to perform an HRA for the HFE (or UA in ATHEANA) of interest.

The intent of such a description is to provide a basic understanding of the progression of events associated with the scenario, and to denote key characteristics that add to its understanding. In other words, the description is intended to identify "what is going on in this scenario," from the perspective of what the operators would see and experience if they were actually involved in the scenario of interest.

Thus, the scenario description needs the following attributes:

- Be well-defined from an operational perspective, in that the key successes and failures of equipment and human responses are clearly delineated.

- Be well-defined from a physics perspective, in that the plant neutronic and thermal-hydraulic responses make sense given the successes and failures during the evolution of the scenario.

- Be realistic from the perspective of accounting for plant-specific features or nuances (to the extent practicable).

- Coincide with operator expectations (because operators typically develop mental models of challenging plant events based on their experience, training, and performance in the plant simulator).

To achieve the above attributes, the analysts should address the following six items in developing the scenario description:

(1)   Provide a brief summary of the assumed initial conditions for the scenario, including the following characteristics (among others):

- mode and power level

- preexisting equipment status (such as failures or other unavailabilities) germane to understanding the progression of the scenario, especially if potentially important for the human action(s) of interest

- any assumed activities in progress (if germane to the scenario progression and/or the human action of interest)

- any other aspects of the pre-scenario plant and human conditions (to the extent relevant to the scenario and its sequence of events)

(2)   Include the initiating event that begins the evolution of the scenario.

(3)   Produce a summary of the sequence of events for the scenario in terms of equipment successes and failures and, if appropriate, other human actions (as might be depicted, for instance, in PRA event trees supported by detailed modeling in fault trees). This should begin with the initiating event and progress primarily in a temporal fashion to the outcome of interest (e.g., core damage), noting the action(s) of interest that the HRA is to assess.

(4)   Include the expected timing of significant plant status changes on the basis of the equipment and human successes and failures postulated in the scenario, and when they are assumed to occur.

(5)   Include expected trajectories, over time, of key parameters (and especially those most relevant to the human action of interest), specifying the status of indications and other cues that are expected as the scenario evolves. Key parameter indications may include the following (among others):

- reactor power and turbine load

- electric power/bus conditions

- status of system flows or isolations (e.g., safety injection, feedwater flow)

- status of support systems such as instrument air, service water, room cooling

- reactor coolant system (RCS) level and pressure

- core heat removal (e.g., T_{avg}, core outlet temperatures, subcooling margin)

- steam generator levels and pressures

- containment pressure and temperature

- radiation indications

- specific equipment conditions (e.g., fluctuating current, high temperature)

- other key parameters addressed in plant-specific emergency operating procedures (EOPs)

(6)   Include assumptions about expected plant behavior, system/equipment/indicator responses, and operator responses relevant to understanding the scenario and its evolution. This is an important part of describing the scenario. As indicated in Section 2, analysts often need to make assumptions about the scenario that are germane to performing the HRA. For example, analysts will typically assume that all instrumentation is available and working properly during the accident sequence unless there is a specific reason to assume otherwise. Such assumptions should be coordinated with the PRA analyst and included in the scenario description.

> In this step, the analysts are to perform the following tasks:
>
> - Provide a description of the scenario relevant to the human action of interest in terms of specific characteristics about the scenario.
>
> - Document the scenario description in a way useful to the performance of other steps in the ATHEANA methodology.

At this point, the description should correspond to that modeled in the PRA so that it represents the evolution that is nominally expected, or is at least a good representative case, given the level of detail provided by the PRA model or other modeling framework being used. So, for instance, if the scenario involves a loss of main feedwater initiator with failure to start of all auxiliary feedwater and with no other equipment failures specifically identified, the scenario description should correspond to these successes and failures.

The description should not, for example, include additional complexities (such as a key instrument failure or additional problems with other equipment, such as a service water pump). These so-called deviations from the nominal case (changes in the context of the scenario) will be examined in Step 6. However, in many cases, analysts make certain assumptions about the evolution of the scenario for the convenience of the PRA. For example, the timing of events may be evaluated on the assumption that all of the failures occur at time zero. This is not necessarily expected, but is assumed for convenience, and often on the basis that it is assumed to be conservatively bounding. Thus, in defining the nominal scenario, analysts need to ensure that assumptions made for convenience do not render the scenario too unrealistic.

Much of the information listed above is routinely collected or generated by analysts in performing an HRA or PRA. Besides the PRA model itself, the following sources may be called upon to provide the scenario description:

- traditional design-basis analyses, such as those covered in plant-specific final safety analysis reports (FSARs)

- plant-specific thermal-hydraulic (T-H) calculations (either existing or performed specifically for the PRA)

- generic engineering analyses for specific issues or initiators

- information sufficient to provide knowledge about the alarms and other cues that will be present, and when (e.g., a listing of alarms and parameter displays by time of occurrence, such as from a simulator run)

- interviews and talk-throughs with operators and trainers (particularly to glean their expectations with regard to the event sequence progression and timing)

- observations (if available) during simulations of the scenario of interest or another scenario with similar characteristics

3.3.2.2   Task 3.2: Document the Description of the Scenario

Regardless of the HRA method used, HRA analysts typically collect and use some or all of the information described above, at varying levels of detail as required by the HRA method. Because other steps in the ATHEANA methodology build upon this original scenario description, and because ATHEANA currently uses an expert elicitation process to quantify the HEPs associated with the HFEs of interest, it is important for the scenario description and its related context to be clear and uniformly available at an appropriate level of detail to enable the experts to assess the importance of various parts of the context, as it relates to performance of the human action(s) of interest.

With this in mind, it is recommended that the scenario description should be in the form of organized notes, models, timelines, text summaries, and any other forms amenable to providing the above information about the scenario.

3.3.3 *Output*

The output of Step 3 consists of a scenario description and its associated context relevant to the human action(s) of interest. The scenario description should include those elements covered under the guidance for this step, and should be documented in a form amenable to communicating what happens in the scenario. This documentation can be a collection of organized notes, models, timelines, text summaries, or whatever else is useful for understanding and communicating the evolution of the scenario.

3.4 Step 4: Define the Corresponding HFE or UA (Including Any EOCs)

3.4.1 *Purpose*

The purpose of this step is for the analysts to identify the human action(s) of interest (if not already determined based on the issue), and to define a corresponding HFE [and associated UA(s), including any EOCs for HRA purposes] that represents the impact of either not performing the required response, or performing an incorrect response. The tasks performed in Step 4 are associated with the requirements covered under HLR-HR-E and HLR-HR-F for post-initiator HRA in the ASME PRA Standard [Ref. 2].

Similar to Step 3, while the guidance that follows is written for a single human action, it is recognized that an issue could involve many human actions of interest for multiple scenarios (e.g., suppose the issue involved investigating the risks associated with key human actions in an entire PRA that examines all types of accident sequences that end with a large early release). The guidance herein would then be applied to all potential actions of interest.

3.4.2 *Guidance*

In this step, the analysts perform the following two tasks:

(1)   Identify the human action of interest relevant to the issue being addressed.

(2)   Define the corresponding HFE and if appropriate, related UAs, including EOCs, that represents the impact of not performing the required response or performing an inappropriate response based on the human action of interest. The level of detail of the definition of the HFE (i.e., HFE or UA, and whether or not an EOC is defined) should be based on what is sufficient to address the issue.

As a reminder, it is likely to be quite useful for the analysts to perform this step and the previous step, Step 3, in parallel. This is because it is natural to consider the description of the scenario and the human action(s) relevant to the scenario, at the same time.

3.4.2.1 Task 4.1: Identify the Human Action of Interest

Compared to other HRA methods that address this task of identifying the human actions of interest (e.g., SHARP1 [Ref. 8]), the ATHEANA methodology provides no unique aspects of this step in performing an HRA. If the definition of the issue is not so exact as to define the human action of interest, the analysts must identify the relevant human action(s) to address the issue.

The activity associated with identifying the human action(s) of interest is to review procedures that are relevant to the issue and to the related scenario described in Step 3, to determine those actions expected to be performed during such a scenario based on procedure directions and operator training. This review should cover the following procedures, as appropriate:

- any relevant EOPs

- abnormal operating procedures (AOPs)

- annunciator response procedures

- system operating procedures

- severe accident management guidelines (SAMGs)

- other procedures, as appropriate (e.g., fire safe shutdown procedures), that are used to direct the desired response

Additionally, in identifying the human action(s) of interest, analysts may find it useful to review actual experience responding to operational disruptions and plant trips.

The identified action(s) should relate to how operators might interface with the systems that are relevant to the scenario of interest. For instance, actions of interest may involve initiating, controlling, operating, isolating, or terminating equipment.

Additionally, analysts should use talk-throughs with plant operating and training staff, as well as simulator observations, to confirm the expected responses for the scenario of interest. This entire process of identifying the human action(s) by reviewing procedures and confirming the expected responses coincides with the requirements covered under HLR-HR-E of the ASME PRA Standard [Ref. 2].

3.4.2.2 Task 4.2: Define the HFE/UA (Including EOCs)

After identifying the human action(s) of interest, the next task is to define the corresponding event that will be evaluated in the HRA and represents the impact of not performing the required response for the given human action(s) or performing an inappropriate action. In doing so, ATHEANA distinguishes between (1) an HFE and its functional effect on the equipment that is to be manipulated, and (2) specific failure modes of the HFE (called UAs). The following definitions are used in ATHEANA:

- HFE (human failure event): a basic event that is typically incorporated into the logic models of a PRA and represents a failure or unavailability of a component, system, or function that is caused by human inaction, or an inappropriate action (virtually identical to the definition in the ASME PRA Standard [Ref. 2]).

- UA (unsafe action): a mode of human failure that results in the HFE and, thus, is a specific action taken or not taken when needed, that results in the failure or unavailability caused by the HFE.

As a beginning point in defining an appropriate HFE for the human action(s) of interest, the analysts should consider the functions with which the operator needs to interact, in responding to the scenario, based on the plant procedures and training. Depending on how the operator, through his or her actions, can affect those functions, candidate HFEs (defined at a high level) to address the human action of interest will become evident. For example, Table 3.4-1 illustrates various ways that operators might influence the functions shown for a scenario involving overcooling. These influences are described in terms of possible classes of HFEs. Depending on the issue to be addressed, one or more of these HFEs can be further defined in more detail, including mentioning the specific system or equipment to be manipulated and when it must be manipulated [e.g., operator fails to close the pressure-operated relief valve (PORV) block valve within 30 minutes], and then be subsequently analyzed in the HRA.

Table 3.4-1. Illustration of Classes of Human Failure Events
Corresponding to Functions Involved in the Scenario of Interest*

Primary Integrity Control	Secondary Pressure Control	Secondary Feed Control	Primary Pressure/Flow Control
• Operator fails to isolate an isolable LOCA in a timely manner (e.g., close a block valve to a stuck-open PORV) • Operator induces a LOCA (e.g., opens a PORV) that induces or enhances a cooldown	• Operator fails to isolate a depressurization condition in a timely manner • Operator isolates when not needed • Operator isolates wrong path/SG • Operator creates an excess steam demand such as opening turbine bypass or atmospheric dump valves	• Operator fails to stop/throttle or properly align feed in • Operator feeds wrong (affected) SG • Operator stops/throttles feed when inappropriate	• Operator does not properly control cooling and throttle or terminate injection to control RCS pressure • Operator trips reactor coolant pumps when not suppose to and/or fails to restore them when desirable
* Note: Some of these are errors of omission (EOOs), and some are errors of commission (EOCs).			

In further developing the definition of the human event to be analyzed in the HRA, in ATHEANA, UAs can be considered as developing an HFE at an even lower level of detail by distinguishing among specific modes of human failure that result in the HFE. For instance, for failure to feed-and-bleed (an HFE), it could be desirable to break down the overall HFE into a "failure to feed" and then separately, a "failure to bleed" (two related but different UAs). Another possible but different breakdown could be "failure to initiate feed-and-bleed" and "failure to adequately control feed-and-bleed once initiated." As part of breaking down failure events into specific UAs, it may also be useful to consider different ways in which a given action might be implemented. For example, an analyst may need or otherwise choose to address a risk-related distinction between an operator inappropriately stopping a pump by turning the pump switch to "off," and inappropriately stopping a pump by pulling-to-lock the pump switch. In both cases, the pump is inappropriately stopped (i.e., the same basic human failure); however, in the first case, the pump may be able to subsequently restart upon a renewed actuation signal, whereas in the second case, it will take another operator action to restart the pump (by the operator having to take the pump control out of the pull-to-lock position).

Figure 3.4-1 demonstrates this concept pictorially by paralleling it with that done for modeling equipment failures. With equipment failures, analysts have to make decisions as to whether the model should depict the specific failure modes of the equipment, or whether it is sufficient to model the equipment failure at a higher level. These decisions often depend on the needs of the analysis. For instance, over the years, typical levels of modeling have evolved, such as often modeling failures of the reactor protection system (RPS) as a high-level "black box," and modeling most other systems and equipment down to specific components and component failure modes. Similarly, the same issue has to be faced with modeling the potential human failures (demonstrated in Figure 3.4-1 in both fault tree and event tree formats). The ATHEANA methodology highlights this issue by emphasizing the distinction between HFEs and the UAs that make up an HFE.

Figure 3.4-1 also illustrates one example of a special form of UA (i.e., adding an EOC as a human failure mode to be treated in the HRA; in this example, "operator prematurely terminates feed-and-bleed"). While the incorporation of EOCs in current PRAs/HRAs is not yet common practice, it is starting to be done in the most current analyses. In this step, ATHEANA specifically calls for the analysts to consider adding EOCs as another form of human failure that may need to be included in assessing the risk related to a failed human response. Oftentimes, serious events have involved EOCs such as terminating safety injection in the Three Mile Island accident, and the decision to launch despite the potential effects of the cold temperature on the solid rocket booster seals of the space shuttle Challenger. Such experiences indicate that it may be important to consider EOCs in HRA.

Table 3.4-2 provides analysts with a list of illustrative UAs (with specific examples of possible forms of the UA) including their classification as an EOO or EOC. In performing Step 4, the analysts should consider such examples as candidates for how to model the human event in the HRA at the UA level if it is necessary or desirable to do so. The UAs correspond to different ways the operator may impact the affected equipment.

In this step, the analysts are to perform the following tasks:

- Identify the human action of interest relevant to the issue being addressed.

- Define the corresponding HFE/UAs (including EOCs) to be evaluated in the HRA.

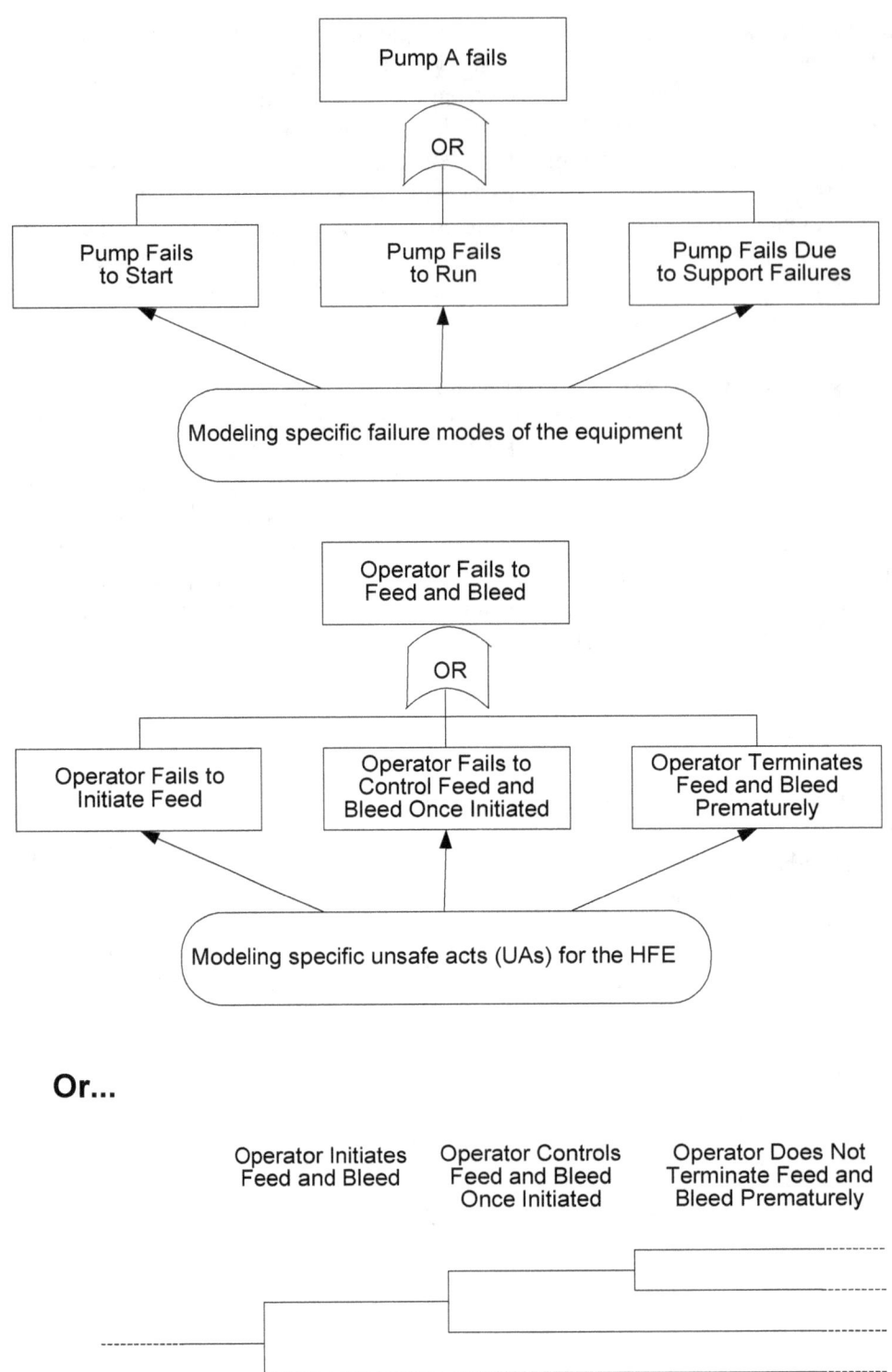

Figure 3.4-1. Depiction of HFEs and UAs as Compared to the Modeling of Equipment Failures and Their Failure Modes

Table 3.4-2. Examples of UAs by PRA Functional Failure Effect

PRA Functional Failure Effect	EOO or EOC?	Examples of Unsafe Actions
Equipment fails to initiate/actuate automatically	EOO	Operator fails to manually initiate/actuate/insert/open/close as a backup
	EOC	Operator prevents automatic initiation/actuation such as by inappropriately removing from automatic control or from armed/standby, pulling to lock the control, disabling automatic control, bypassing auto signals
Equipment fails to continue to operate for mission time	EOC	Operator prevents continuation of equipment operation such as by inappropriately terminating, isolating, realigning, diverting the output/flow, depleting the output/resources, stopping and pulling to lock the control, actuating automatic isolation or realignment signals
Equipment fails to be manually initiated/actuated when required	EOO	Operator fails to manually initiate/actuate/insert/open/close when required or manually initiates but too late
	EOC	Operator inappropriately initiates/actuates too soon or when not needed
Equipment fails to be controlled or operated as required	EOO	Operator fails to operate or control resulting in under/overfeeding, under/overcooling, under/overpressure, reactivity increase/decrease
	EOC	Operator inappropriately operates or controls resulting in the above
Equipment fails to maintain desired status	EOO	Operator fails to maintain equipment status such as by failing to maintain integrity or keep alignment
	EOC	Operator inappropriately changes equipment status such as by inappropriately breaching integrity or realigning
Equipment fails to stop automatically	EOO	Operator fails to manually stop/terminate/withdraw/open/close as a backup
	EOC	Operator prevents automatic stop of equipment such as by inappropriately removing from automatic control, disabling automatic control, bypassing stop/isolation signals
Equipment fails to be stopped manually when required	EOO	Operator fails to manually stop/terminate/withdraw/ open/close when required or manually stops too late
	EOC	Operator inappropriately stops too soon or when not needed
Equipment fails to remain stopped for required duration	EOC	Operator inappropriately restarts

Deciding whether it is desirable for the HRA to model/analyze a human failure at the more general HFE level or at the more distinctive UA level, and whether or not an EOC form of failure should be included, depends on the issue being addressed and the anticipated uses of the HRA results. Modeling and analyzing at the UA level provides the means to explicitly investigate the potential impact of different UAs, including EOCs, on the plant response and other human actions. The importance of considering different impacts is addressed, for instance, in Section 3.1.2.2 of the SHARP1 methodology [Ref. 8]. However, as also delineated in that same reference, it is not always practical or even necessary to breakdown a human action into different specific failure modes and underlying causes. If the issue does not specify a particular level of resolution of the analysis results from the HRA, such as addressing a specific human failure mode, the following are offered as guidelines that the HRA analyst may when deciding whether to model at the HFE or UA level:

- If the performance influencing factors that could lead to or otherwise most affect the likelihood of the possible UAs (including EOCs) that might make up the HFE are identified as being significantly different, consider modeling at the UA level if it is desirable or necessary for the HRA results to reflect these differences.

- If the perceived/anticipated error rates, including any recovery potential, are likely to be very different among the UAs (including EOCs) comprising the HFE, consider modeling at the UA level if it is desirable or necessary for the HRA results to reflect these individual error rates.

- If there is a perceived significant dependency between a particular UA (including EOCs) that is associated with the HFE and some other human failure modeled in the PRA (either upstream or downstream in the chain of events depicted by the PRA sequence), consider modeling at the UA level. This may be particularly important when the various UAs would lead to differences in the subsequent scenario development. By breaking the HFE into UAs (including EOCs), the specific dependency can be modeled more appropriately and explicitly.

This task corresponds most directly to the HR-F1 supporting requirement in the ASME PRA Standard [Ref. 2], which covers defining HFEs. Note that the associated ASME supporting requirement HR-F2, which addresses completion of the HFE definition, is addressed in ATHEANA by (1) performing this task in conjunction with Task 4.1 (immediately above) concerning the action(s) of interest, and (2) while considering the information about the scenario of interest (e.g., the timing and availability of cues, time available to take action) in order to complete the definition of an HFE/UA (including EOCs).

3.4.3 *Output*

The output of Step 4 consists of a definition of the human event to be addressed in the HRA, in terms of an HFE or one or more UAs, including possible EOCs, as necessary or desirable to sufficiently address the issue. The HFE/UA (including EOCs) is to represent the impact of not performing the required response for the human action(s) of interest or performing an inappropriate related action.

3.5 Step 5: Assessing Human Performance Relevant Information and Characterizing Factors That Could Lead to Potential Vulnerabilities

3.5.1 *Purpose*

The purpose of this step is to identify and characterize factors (e.g., PSFs) that could contribute to crew performance in responding to the various accident scenarios, with the main interest being to identify factors that could create potential vulnerabilities in the crew's ability to respond to the scenario(s) of interest and increase the likelihood of the HFEs or UAs identified in Step 4. Thus, this step provides information that is critical to modeling the context of the HFEs/UAs, quantifying the events, and searching for important deviation scenarios. This step corresponds to collecting and assessing the qualitative aspects (e.g., perform an analysis, evaluate PSFs, consider timing, assess dependencies) called out under the HLR-HR-G requirement and its supporting requirements in the ASME PRA Standard [Ref. 2]. These assessments will be used as inputs to the quantification process (Step 8), where the HEPs for the HFEs/UAs are estimated.

This step bridges the initial efforts of the HRA process in Steps 1–4 and the subsequent analysis. Toward that end, the analysts organize and familiarize themselves with information relevant to the HFEs/UAs of interest, in the context of the described nominal scenarios for the related PRA accident sequences, for easy access in subsequent steps of the analysis. The initial focus is on identifying and understanding the factors that could influence successful performance of each of the HFEs/UAs in each of the given nominal scenarios for the identified PRA accident sequences. This information will then be used to support the quantification of the HFEs/UAs in the nominal scenarios during Step 8.

During the deviation analysis to be performed in Step 6, analysts examine the various nominal scenarios for potential variations in plant or situational conditions that might capitalize on the vulnerabilities identified in this step (Step 5), or might create new vulnerabilities and lead to a strong EFC. For example, the way the crew members interact and work as a team may be generally effective for the nominal scenario, but their modus operandi may become a vulnerability under somewhat different conditions. Similarly, formal procedures may work well in most cases, but may not match well with certain scenario evolutions or physical conditions in other cases. Thus, another purpose of Step 5 is to understand the factors that are relevant to each accident scenario, at least well enough to allow the analysts to investigate (in Step 6) whether realistic variations in conditions could trigger them into becoming vulnerabilities and create challenging conditions for the operating crew. As previously discussed, certain combinations of plant conditions and PSFs can create an EFC and the resulting situations observed in serious accidents.

The following subsections discuss activities to perform and several areas to explore to help understand how an operating crew might respond to the initiating event in the nominal scenario and to identify potential problem areas related to their response. The analysts begin the process of looking for inherent characteristics of the operating crews (e.g., team dynamics, strategies for implementing procedures) or the crew's knowledge base (informal rules, expectations or biases based on training, learned response tendencies, etc.), which could contribute (either positively or negatively) to the HFE/UA of concern. In addition, the analysts examine the factors that can directly affect what the crews will do, such as formal procedures and written rules, the human-machine interface, the timing of the scenarios, environmental conditions, and so forth.

In general, although some of the factors or characteristics addressed here have not always been included in the lists of PSFs considered in an HRA, and although the various factors may manifest their effects in different ways, they all have the potential to create vulnerabilities if the accident scenarios evolve in certain ways. In other words, they are all PSFs, in the sense that they can either facilitate or hinder crew performance, depending on plant conditions and the way the accident scenario evolves. Thus, in keeping with traditional HRA terminology, the influences are all discussed as PSFs that analysts must consider when examining what the crews are likely to do in a given accident scenario, and identifying potential vulnerabilities.

It is important to reemphasize that understanding the various PSFs (as described below) is relevant to analyzing and quantifying events in the nominal scenarios that have traditionally been quantified in the PRA, but it also forms the basis for addressing deviation scenarios in Step 6. The perspective is that the broad range of factors that are considered here and carried forward to quantifying the HFE/UA for the nominal case, and eventually identifying potential deviation scenarios and quantifying the HFEs/UAs for those scenarios, provides the basis for realistic assessing the probability of failure of human events in PRA accident scenarios.

Note that HRA and PRA analysts can use interviews with plant personnel (e.g., trainers and operators), to search for and document much of the information needed for this step of the analysis. However, for the actual evaluation of the impact of PSFs and performance dependencies (Task 5.3 below), it is recommended that the HRA team should be assembled and the influences should be discussed as a group, with the HRA analyst documenting the results for later use.

3.5.2   *Guidance*

This step requires three major tasks to be performed:

(1)   Using the PRA-specified sequence of events and time line developed for each of the nominal scenarios and HFEs/UAs in Steps 3 and 4, (a) structure the scenario information into time frames as needed to help identify inherent difficulties in the required response (e.g., very limited response time), and (b) prepare to evaluate the scenarios with respect to the effect of PSFs on likely crew performance.

(2)   Collect or develop the plant-specific information relevant to understanding the role of the PSFs and how they might bear on the scenario and responses being examined (e.g., effect of training schedules on the various scenarios). This task also addresses the use of simulator exercises to obtain important information.

(3)   Review each of the PSFs against each accident scenario and relevant HFEs/UAs to identify potential vulnerabilities and to assess their potential positive and negative influences on performance in the nominal case.[7] At this point, analysts will be identifying those PSFs

---

[7]   Note that an identified negative effect will generally indicate a vulnerability. For example, the degree of regular training on a given nominal scenario may be limited, which could increase the likelihood of failure for an HFE/UA in the nominal case. Thus, level of training could be seen as a negative influence and a vulnerability, which could also become more problematic for particular deviation scenarios identified in Step 6. Nonetheless, analysts may identify potential vulnerabilities in Step 5 that would not likely influence performance in the nominal case, but could influence performance if conditions change. For example, certain identified differences between plant crews (e.g., aggressiveness in responding) may not be expected to affect success given a long time-frame is assumed for the nominal scenarios, but the crew differences might create a vulnerability if the scenario moved at a faster rate as a result of a reasonable likely concurrent failure. In this case, the vulnerability would be relevant for consideration during the deviation analysis, but would not be a negative influence for the nominal case.

that are likely to have effects, with the final strength of their impact on successful performance assessed during the quantification step. The assessment of potential dependencies between events in the scenarios will also be initially evaluated in this task, because crew understanding and decisions made early in a scenario could affect their ability to respond later.

3.5.2.1 Task 5.1: Structure the Scenario and HFE/UA Information from Steps 3 and 4 To Support the Evaluation of PSFs on Crew Performance

A review of the PRA-specified sequence of events and time line developed for each of the nominal scenarios and HFEs/UAs in Steps 3 and 4, will generally reveal natural time frames for a given scenario with respect to plant behavior, plant symptoms, system response, and operator response. These aspects usually align with the following phases of the scenario:

- initial conditions or pre-trip scenario

- initiator and nearly simultaneous events

- early equipment initiation and operator response

- stabilization

- long-term equipment and operator response

Recasting the scenario description and time line into a concise presentation of these natural time frames can prove helpful, exposing the bases for many of the equipment success criteria and clearly identifying periods of minimal/maximal vulnerability to inappropriate human intervention (e.g., phases in the scenario where delayed crew response could cause problems, or where it might be difficult to recover from inappropriate actions, as well as times of very high workload or simultaneous activities that may affect proper crew response). Examples of such time frame analyses are presented in Table 3.5-1. The analysts should note and/or document any potentially difficult phases of the scenario and the potential operator vulnerabilities. In the next steps, the developed time frames and general time line information will be considered in conjunction with PSFs, such as the plant-specific operating procedures, to identify potential vulnerabilities and important influences on performance.

Table 3.5-1. Relevant Time Frames for Two Example Scenarios

Time Frame	Loss of Main Feedwater (MFW) Scenario		Large LOCA Scenario	
	Major Occurrences	**Influences on/by Operators**	**Major Occurrences**	**Influences on/by Operators**
Initial conditions	Steady-state, 100% power No previous dependent events in nominal scenario	Routine conditions; nothing to focus attention	Steady-state, 100% power No previous dependent events in nominal scenario	Routine conditions; nothing to focus attention
Initiator or/ simultaneous events	Loss of MFW Reactor scram or turbine trip	Operators may identify MFW problems and manually trip the plant.	Reactor power prompt drop Pressure drops below safety injection (SI) initiation point	These events are over before the operator even recognizes what is happening
Early equipment initiation and operator response	0–2 minutes Auxiliary feedwater (AFW) start SG pressure control per blowdown Other auto equipment responses	Operators verify initial responses per EOPs; particularly, AFW start in this case. Operators may even manually start AFW before it auto starts.	0–20 seconds Break flow is complete Pressure drops to essentially zero Containment pressure has peaked and is falling ECCS flow begins Accumulator flow occurs	During this time frame the operator is checking parameters and ensuring that appropriate standby equipment has started. Some early decisions in the EOPs may have occurred.
Stabilization phase	2 minutes – 1 hour Heat sink restored (SG levels and pressure) Plant conditions restabilize Some throttling and shutting down of equipment (e.g., AFW) begins	Operators likely to throttle and even shut down some AFW pumps to avoid overcooling or respond to lack of cooling (and enter other EOPs) if heat sink apparently not restoring. Perform other actions as necessary (e.g., pressurizer heater on or off) to keep plant stabilized.	1–3 minutes Core reflood begins at about 30 seconds and has reached stable conditions Fuel temperatures have peaked and are falling	During this time, the operators have moved into the LOCA EOP and have passed a number of decision points.
Long-term equipment and operator response	>1 hour Unnecessary equipment shutdown Achieve hot or cold shutdown	Operator shuts down unnecessary equipment and transitions plant to hot or cold shutdown.	Isolation of the accumulators Shift to cold leg recirculation cooling Shift to hot leg recirculation cooling Repair and recovery	During the 20 minutes until switchover to cold leg recirculation cooling, the operators are occupied with confirmatory steps in the EOPs. Any complications beyond the nominal scenarios can affect their performance. This longer time frame extends to days and months. There are no critical operations concerned with the nominal scenario. Problems during this phase would be the concern of a low-power and shutdown PRA.

3.5.2.2 Task 5.2: Collect or Develop the Plant-Specific Information Relevant to Understanding the Role of the PSFs and How They Might Bear on the Scenario Being Examined

This section describes a set of PSFs and some key characteristics to consider when evaluating crew performance in a given scenario, relative to a particular HFE/UA. In addition, these descriptions include potential interactions among the PSFs, which analysts should examine when assessing the impact of the PSFs. In order to be able to evaluate the PSFs in light of the relevant scenarios, the analysts will need to collect or develop certain plant- and scenario- specific information. These PSFs should be considered in terms of their potential effects on both the diagnosis of needed actions and the execution of those actions, including both within and outside the control room. Analysts should review the PSF descriptions and collect the information that will be needed to evaluate their influences on the various scenarios and specific HFEs/UAs being modeled. After collecting and documenting all of the needed information, analysts can proceed to the evaluation process (Task 5.3 of this step); however, there tends to be significant iteration between Tasks 5.2 and 5.3.

In this task, analysts also consider the potential aleatory influences that could affect eventual estimates of HEPs (e.g., random instrument failures, important variations in crew characteristics, weather variations when actions must be conducted outside, possible recurring workarounds). That is, to the extent aspects of the PSFs vary randomly (aleatory factors), variations in the HEP could be expected, and could contribute to the range associated with probability estimates depending on the specific context.

The reader should note that the goal of the following PSF descriptions is to provide analysts with guidance regarding the types of influences they should consider in evaluating human performance. Many of the PSF descriptions are somewhat complex, and the factors influencing human performance are equally complex and overlapping and cannot always be defined in a simplistic way. Thus, analysts should read these PSF descriptions as guidance regarding the plant- and scenario-specific information they need to collect and consider in order to produce an analysis that is as realistic and complete as possible (or as specified by the goals of the given analysis).

Descriptions of PSFs

1. *Applicability and Suitability of Training/Experience*

For both in-control room and local actions (including diagnosis and response execution), the applicability and suitability of training/experience is an important factor in assessing operator performance. In general, in nuclear power plants, operators can be considered to be trained at some minimum level to perform their required tasks.

However, from an HRA perspective, the degree of familiarity with the type of sequences to be addressed and actions to be performed, can provide either a negative or positive influence that should be considered to assess the likelihood of operator success. In cases where the type of PRA sequence being examined or the actions to be taken are not periodically addressed in training (in classroom sessions or simulations every 1–2 years or more often, for example), or the actions are not performed as part of the operators' normal experience or on-the-job duties, this factor should be treated as a negative influence, while the converse would result in a positive influence on overall operator performance.

One should also attempt to identify systematic training biases that may have a positive or negative effect on operator performance. For example, training guidance in a pressurized-water reactor (PWR) may induce a reluctance to use "feed-and-bleed" in a situation where steam generator feed is expected to be recovered. Other training experience may suggest that operators are allowed to take certain actions before the procedural steps calling for those actions are reached, if the operators are sure the actions are needed. Such training biases could cause hesitation and result in higher HEPs for the desired actions, as in the first case above, or may be a positive influence as in the case of proactively taking obvious actions.

Related to the discussion of biases, Reason [Ref. 9] identified particular human information processing heuristics[8] or strategies (e.g., pattern matching rather than more systematic problem solving) that have particularly powerful effects on people when they must make decisions about events. Such heuristics, in conjunction with certain biases, can affect the kinds of choices operators make under abnormal conditions. The three most common biases associated with these heuristics that may relate to control room operations during abnormal conditions are as follows:

- Recency: operators are biased to recall events that occurred recently or were the subject of recent operational experience, training, or discussions (but note that recency biases may be particularly transitory; see the discussion in the next paragraph)

- Frequency: operators are biased to recall events that they frequently encountered in situations that appear (even superficially) similar to the scenario currently being encountered

- Similarity: operators are biased to recall events with characteristics that are (even superficially) similar to the current scenario.

Such biases or expectations can lead crews to inappropriately interpret a somewhat unexpected scenario in ways that make the current scenario fit with past experience, which in turn can lead the crew to make incorrect choices. Identification of these biases can help to identify, for example, the more likely incorrect situation assessments, where operating crews may overlook or become preoccupied with particular parameter indications. However, the reader should note that such biases may only become vulnerabilities if somewhat deviant scenarios are feasible, and if they evolve in a way that capitalizes on the biases. Thus, prominent biases on the part of the crew (or even particular crew positions), should be documented and retained for later consideration as potential vulnerabilities during the deviation analysis in Step 6.

Moreover, the analyst will have to decide whether such biases qualify as "ingrained" factors of the crew before it makes sense to model their contributions to potential HFEs/UAs. For example, recency effects are usually transitory, but the events at Three Mile Island in 1979 may have remained "recent" in the minds of PWR crews for many years. Similarly, if crews experience scenarios for particular initiating events in very similar ways each time they are run in the training simulator, those simulations are more likely to create an ingrained expectation for how the scenario will evolve than they would be if the simulations included regular and sometimes dramatic variations in the conditions. Thus, analysts should search for and note or document ingrained biases to be evaluated against the nominal and deviation cases, to see if they might contribute to inappropriate responses. Interviews and discussions with operators can help to identify scenarios with similar signatures that might be confused, and can help to determine the types of scenarios with which the crew is most familiar as a result of frequent training.

---

[8]   A *heuristic* is a mental shortcut in recognizing a situation and taking an action. Heuristics normally allow people to quickly select the most plausible choices before those that are less plausible.

The potential negative effects of familiarity and similarity biases were some of the driving concerns underlying the development of the confusion matrix approach [Refs. 10 and 11] for identifying potential UAs and why they might occur. The confusion matrix can be another useful tool for analysts to use in the process of identifying potential vulnerabilities.

Thus, it is incumbent on the analyst to ensure that training and/or experience is relevant to the PRA sequence and desired actions. The more it can be argued that the training is current, is similar to the actual event, is sufficiently varied to represent differences in the way the event may evolve, and is conducted frequently enough to demonstrate proficiency on a periodic basis, the more positive this factor can become. By contrast, if there is little or no training/experience, or there are potentially negative training biases for the PRA sequence being examined, this factor should be considered to have a negative influence. Potentially inadequate training and identified biases should also be carried forward as potential vulnerabilities for consideration during the deviation analysis.

2.   *Suitability of Relevant Procedures and Administrative Controls*

For both in-control room and local actions, the suitability of relevant procedures and administrative controls is an important factor in assessing operator performance. Similar to training, procedures generally exist to cover most types of sequences and operator actions.

However, from an HRA perspective, the degree to which procedures clearly and unambiguously cover the types of sequences to be addressed and actions to be performed, can provide either a negative or positive influence that should be considered to assess the likelihood of operator success. In general, this PSF should be considered to be adequate (or even a positive influence) unless the procedures have the following (or similar) characteristics related to the desired actions for the sequences of interest.

- ambiguous, unclear, or non-detailed steps for the desired actions in the context of the sequence of interest

- situations where the operators are likely to have trouble identifying a way to proceed forward through the procedure

- requirements to rely on considerable memory

- situations in which operators must perform calculations or make other manual adjustments (especially time-sensitive situations)

- situations for which there is no procedure, or the procedure is likely not to be available, especially when taking local actions "in the heat of the scenario" and when it cannot be argued that the desired task is simple and a "skill of the craft" or it is an automatic or memorized activity on which the crew is trained and has routine experience

- the procedures contain "double negatives" (these should be evaluated to determine whether certain circumstances could make the procedures particularly confusing)

To support the assessment of this PSF, two key activities are suggested. First, talk-throughs of procedures with operations and training staff (in the context of the scenario being examined) can be helpful in revealing difficulties or ease of use, considering the associated training that the operators receive and the ways the operators interpret the procedures. Second, a flowchart or logic diagram format representing each procedure and the interconnections between them is suggested as an aid to analysts. These flowcharts or logic diagrams should distinguish between the procedure steps in which decisions are made and those where actions, monitoring, or verification is performed. Such simplified diagrams highlight the following considerations:

- related text that denotes the entry conditions for the procedure
- locations of branch points from the most applicable procedure to other procedures
- specific steps that call for starting, stopping, or otherwise affecting equipment that is particularly germane to the scenario
- where a major reconfiguration of equipment is called for

Appendix A to this user's guide presents an example flowchart of emergency operating procedures. In addition, several examples of flowcharts of emergency operating procedures are presented in Appendices B–E of NUREG-1624, Rev.1 [Ref. 1].

The EOPs or other formal rules define the responses operators are expected to take, depending on the scenario progression. However, the entry, branch, and decision points in the procedures could be particularly vulnerable to operator errors that might cause an incorrect procedure to be entered or equipment to be inappropriately shut down or reconfigured. Therefore, at each decision point (or where otherwise deemed beneficial), summaries of actions to be taken, the potential for ambiguity, and a judgment on the significance of taking inappropriate action should be noted and will provide clues to possible pitfalls under certain circumstances. This information (potential vulnerabilities) will also be used during the deviation analysis to help identify where reasonable, but unusual circumstances might create mismatches between the evolving conditions and the procedures.

All competent analysts examine plant procedures and consider their impact on operations. Other existing methods [see, for example, Refs. 4–8, 10, and 11]) also encourage a rigorous review of procedures for potential problems with respect to specific scenarios.

3. *Operator Action Tendencies and Informal Rules*

Operator action tendencies are based on the formal emergency and abnormal operating procedures and related training that is received, as well as informal practices/rules that are part of the operator psyche. These tendencies and informal rules can strongly influence what crews will do under given circumstances. In considering operator tendencies, analysts should identify those that may be either positive or negative influences with respect to the HFEs/UAs of interest under the conditions of the scenarios being addressed. In addition, analysts should identify any informal rules in the plant that may be relevant as possible contributing factors to the occurrence of the HFEs/UAs. For example, an informal rule may exist among the operating staff that a certain indicator should not be trusted because it often sticks and, thus, reads incorrectly during dynamic situations (essentially a permanent workaround). If the analysts identify a way that following this or other informal rule could contribute to the likelihood of an HFE/UA, the circumstances associated with the HFE/UA should be examined to see if the informal rule might be a contributing positive or negative influence. Tables 3.5-2a, 3.5-2b, and 3.5-3 identify examples of tendencies and informal rules, and most are generally applicable for many different types of scenarios in different HRAs.

Non-procedure based tendencies and informal rules (e.g, based on training, experience, and general "plant wisdom") can be obtained through interviews with operators, trainers, and in some cases, management personnel. It is particularly useful to observe and discuss simulator exercises with crews and trainers to glean information about tendencies and informal rules (additional information on the use of simulator exercises is provided later in this section). This type of information can be revealed all the way through the HRA analysis, including during quantification, as different scenarios are discussed with plant personnel.

Analysts should carefully document identified tendencies and informal rules and the circumstances that can elicit them for use during the deviation analysis. It is under unusual or unexpected conditions where tendencies and informal rules might result in misapplication and have a significant impact on performance.

Table 3.5-2a. Summary of Operator Action Tendencies (PWRs)

Key Functional Parameter(s)	Off-Normal Condition[a]	Operator Action Tendencies[b]
Plant Level: Reactor power	Too high or increasing	Rods in or Emergency borate (inject)
Turbine/generator load	Not tripped	Trip / Run back /close main steam valves
Key Supports: Electric power	Partial or total loss	Restore (use emergency diesels if necessary) or realign
Cooling water systems	Partial or total loss	Restore/realign/augment
Instrument air	Partial or total loss	Restore or realign
Reactor Coolant System (RCS) (primary): Pressurizer (RCS) level	Too low or decreasing	More RCS injection or less letdown
	Too high or increasing	Less/stop injection or more letdown
Pressurizer (RCS) pressure	Too low or decreasing	More RCS injection / isolate possible LOCA paths / stop pressurizer sprays and turn on heaters / decrease cooldown
	Too high or increasing	Turn on pressurizer sprays and turn off heaters / increase cooldown / provide relief with pressure-operated relief valves (PORVs) or vents
Core heat removal (e.g., T_{avg}, core outlet temps, subcooling)	Too low or decreasing (insufficient)	Increase RCS forced flow (unless voiding evident) / more RCS injection / increase cooldown
	Too high or increasing (overcooling)	Decrease RCS forced flow / less/stop injection / close any open PORVs/vents / decrease cooldown

Key Functional Parameter(s)	Off-Normal Condition[a]	Operator Action Tendencies[b]
Steam Generators (SG) (secondary): SG Level	Too low or decreasing	More SG feed (i.e., increase cooldown) / use feed-and-bleed
	Too high or increasing	Less SG feed (i.e., decrease cooldown) / possible isolation of main steam
SG Pressure	Too low or decreasing	Decrease steam dump (i.e., decrease cooldown) / isolate (especially in the event of high radiation indicative of tube rupture)
	Too high or increasing	Increase steam dump or provide main steam relief (i.e., increase cooldown)
Containment: Containment pressure	Too high or increasing	Increase fan cooling / isolate containment / containment spray
Containment temperature	Too high or increasing	Increase fan cooling / isolate containment / containment spray
Radiation	Indicating	Isolate source or area
Ventilation	Too little or rising temperature	Regain / open doors/ use portable equipment
Other: Equipment condition	Signs of imminent damage (vibration, fluctuating current, high temperature)	Shut down or isolate

[a] This is defined relative to what is expected at the time in the scenario when the operator is responding to the functional parameter of interest. Note that the operator may respond to a parameter early in the event and again later in the event and so forth. The expected absolute reading or trend of the parameter could be different for the early and later responses. The off-normal condition is defined relative to each expectation at each time.

[b] It is recognized that the specific actions will depend on the absolute reading and rate of change in the parameter and the specific procedural guidance for the conditions observed. These are, however, the typical types of actions that are called out to be performed, depending on the specific circumstances.

Table 3.5-2b. Summary of Operator Action Tendencies (BWRs)

Key Functional Parameter(s)	Off-Normal Condition[a]	Operator Action Tendencies[b]
Plant Level: Reactor power	Too high or increasing	Rods in / emergency borate/ level-power control
Turbine or generator load	Not tripped	Trip / Run back / close steam valves
Key Supports: Electric power	Partial or total loss	Restore (use emergency diesels if necessary)/realign
Instrument air	Partial or total loss	Restore or realign

Key Functional Parameter(s)	Off-Normal Condition[a]	Operator Action Tendencies[b]
Cooling water systems	Partial or total loss	Restore/realign/augment
Reactor Pressure Vessel: Level	Too low or decreasing	More vessel injection / depressurize /vessel flooding/ isolate containment / containment flooding
	Too high or increasing	Reduce feedwater or less-stop injection
Pressure	Too high or increasing	Provide relief [turbine bypass, safety relief valves (SRVs), etc.]
Containment: Suppression pool temp.	Too high or increasing	Suppression pool cooling sprays or depressurize
Suppression pool level	Too high or increasing	Use pool drains / terminate external injection / depressurize
	Too low or decreasing	Provide pool makeup or depressurize
Drywell pressure	Too high or increasing	Isolate LOCA and containment / drywell spray / venting / depressurize
Drywell temperature	Too high or increasing	Increase drywell cooling / drywell spray / depressurize
Radiation	Indicating	Isolate source/area / depressurize
Ventilation	Too little and/or rising temp	Regain / open doors/ use portable equipment
Other: Equipment condition	Signs of imminent damage (vibration, fluctuating current, high temperature)	Shutdown / isolate

[a] This is defined relative to what is expected at the time in the scenario when the operator is responding to the functional parameter of interest. Note that the operator may respond to a parameter early in the event, and again later in the event, and so forth. The expected absolute reading or trend of the parameter could be different for the early and later responses. The off-normal condition is defined relative to each expectation at each time.

[b] It is recognized that the specific actions will depend on the absolute reading and rate of change in the parameter and the specific procedural guidance for the conditions observed. These are, however, the typical types of actions that are called out to be performed depending on the specific circumstances.

Table 3.5-3. Examples of Informal Rules Used by Operators

How Operators Use Rules	Informal[a]	
	Training	**Other Sources of Informal Rules**
Plant Interventions		

Selection and justification of unsafe action(s)	Keep core covered
	Always follow your procedures
	Don't go solid in pressurizer
		Good Practice
		    Protect pumps (e.g., stop if no lube oil pressure, no cooling, runout, deadheaded, cycling)
		Old Practice
		    Safety injection (SI) on low pressurizer level
		Folklore
		    A good operator always beats autoactuation
		    Never feed water into an overheated vessel
		Conflict
		    Alternatives have negative consequences
		    Success seems imminent

Information Processing		
Monitoring[b] (i.e., what indications to monitor, when to monitor, etc.)	Which instruments to use Which (and in what order) to respond to alarms Check redundant indications (especially alarmed conditions)	Experience Which instruments to use (may not be all that are available)

Interpretation (part of situation assessment)	Believe your indications	Good practice
		    Question diagnoses (e.g., if unexpected response, restore your last action)
		Experience (plant-specific)
		    Some indications are more reliable than others.
		    Some indications always give false readings.
		    Recent history of plant/equipment/instrument performance

Understanding Plant Conditions and Configurations		

Equipment status	Indications of performance.
	Believe your tagout system
		Folklore
		    Pumps in runout overspeed
		    Multiple failures in one system are not possible

Instruments/indications	Instruments are very reliable	Folklore
		    Indication readings correspond directly with actual plant state or behavior
		    Indications are independent

[a] Including training, guidance for good operating practice, old practice (i.e., previous operating practice), experience, invented rules of thumb (referred to as folklore).

[b] Including both data-driven and knowledge-driven monitoring.

4.   *Availability and Clarity of Instrumentation (Cues to Take Actions and Confirm Expected Plant Response)*

For both in-control room and local actions, the availability and clarity of instrumentation is an important factor because operators, other than for immediate and memorized response actions, take actions based on diagnostic indications and look for expected plant responses to dictate follow-on actions. For in-control room actions, typical nuclear plant control rooms have sufficient redundancy and diversity for most important plant parameters. Consequently, most HRA methods inherently assume that adequate instrumentation typically exists. Nonetheless, this should be verified by looking for the following characteristics that could make this a negative PSF, particularly in situations where there is little redundancy in the instrumentation associated with the action(s) of interest:

*   The key instrumentation associated with an action is adversely affected by the initiating event or subsequent equipment failure (e.g., loss of DC power causing loss of some indications, spurious or failed as a result of a hot short from a fire).

*   The key instrumentation is not readily available and may not be typically scanned (such as on an obscure back panel).

*   The instrumentation could be misunderstood or may be ambiguous because it is not a direct indication of the equipment status (e.g., PORV position is really the position of the solenoid valve and not the PORV itself).

*   The instrumentation is operating under conditions for which it is not appropriate (e.g., calibrated for normal power conditions rather than shutdown conditions).

*   There are so many simultaneous changing indications and alarms or the indication is so subtle, particularly when the time to act is short, it may be difficult to "see and pick out" the important cue in time (e.g., a changing open-close light for a valve without a concurrent alarm or other indication, finding one alarm light among hundreds).

The above also applies to local actions outside the control room, recognizing that in some situations, less instrumentation may exist (e.g., only one division of instrumentation and limited device actuators on the remote shutdown panel). However, on the positive side, local action indications often can include actual/physical observation of the equipment (e.g., pump is running, valve stem shows it is closed), which compensates for any lack of other indicators or alarms.

It is incumbent on the analyst to determine whether adequate instrumentation is available and clear so that the operators will know the status of the plant and when certain actions need to be taken. If this is demonstrated, this PSF would be positive. Task analysis will often facilitate determining whether the instrumentation is adequate.

Given the importance of adequate indications and cues for a needed diagnosis and response, during the assessment of such information for the nominal case, analysts should consider the potential impacts on performance if certain instruments failed or were not functioning exactly right (e.g., if multiple inputs were required for a particular indication and some of that information was faulted in some way. If, for example, because of a lack of redundancy or difficulty in detecting that a particular instrument is not functioning correctly (e.g., for subcooling), the potential for a UA would be likely to increase significantly, then such information (a potential vulnerability) should be collected for later consideration during the deviation analysis, where a range of contexts other than the nominal will be considered.

5.   *Time Available and Time Required to Complete the Act, Including the Impact of Concurrent and Competing Activities*

This can be an important influence for both in-control room and local actions because, clearly, if there is not enough or barely enough time to act, the estimated HEP is expected to be quite high. Conversely, if the time available far exceeds the time required and there are not multiple competing tasks, the estimated HEP is not expected to be strongly influenced by this factor.

It is important that analysts consider the time available and the time needed to perform the action in concert with many of the other PSFs and the demands of the sequence. This is because the T-H inputs (e.g., time to steam generator dryout, time to start uncovering the core), while important, are not the only influences. (Note, it is best if the T-H influences are derived from plant-specific or similar analyses, rather than simple judgments).

The time to perform the act, in particular, is a function of the following factors:

*   number of available staff

*   clarity and repetitiveness of the cues that the action needs to be performed

*   the human-system interface (HSI, discussed later)

*   complexity involved (discussed later)

*   need to get special tools or clothing (discussed later)

*   consideration of diversions and other concurrent requirements (discussed later)

*   where in the procedures the steps for the action of interest are called out

*   crew characteristics such as whether the crews are generally aggressive or slow and methodical in getting through the procedural steps (discussed later)

*   other potential "time sinks"

Clearly, there is judgment involved; however, as described here, it is not as simple as watching an operator perform an action in ideal conditions with a stop watch to determine the time required to perform the act. Only when the sequence context is considered holistically with the interfacing PSFs that have been mentioned here, can more meaningful times be estimated. Hence, especially for complex actions and/or situations, walkdowns and simulations can be helpful in ensuring overly optimistic times have not been estimated.

Also, it should be emphasized that when feasible, observation of simulator exercises can be used to obtain information on the typical times it takes crews to get through various procedure steps, respond to events, and deal with unexpected failures or distractions. While simulator data cannot always be used directly (see the section below on simulator exercises), the knowledge can be a useful input into the role that time plays in particular scenarios.

6.   *Complexity of the Required Diagnosis and Response, the Need for Special Sequencing, and the Familiarity of the Situation*

This factor attempts to measure the overall complexity involved in the situation and action of interest (e.g., the same operator must perform many steps in rapid succession vs. one simple skill-of-the-craft action). Many of the other PSFs bear on the overall complexity, such as the need to decipher numerous indications and alarms, the presence of many and complicated steps in a procedure, or poor HSI. Nonetheless, this factor should also capture measures such as the ambiguity associated with assessing the situation or executing the task, the degree of mental effort or knowledge involved, whether it is a multi-variable or single-variable associated task, whether special sequencing or coordination is required in order for the action to be successful (especially if it involves multiple persons in different locations), or whether the activity may require very sensitive and careful manipulations by the operator. The more these measures describe an overall complex situation, this PSF should be found to be a negative influence. To the extent these measures suggest a simple, straightforward, unambiguous process (or one that the crew or individual is very familiar with and skilled at performing), this factor should be found to be nominal or even ideal (i.e., positive influence).

7.   *Workload, Time Pressure, and Stress*

Although workload, time pressure, and stress are often associated with complexity and can certainly contribute to perceived complexity, the emphasis here is on the amount of work a crew or individual has to accomplish in the time available (e.g., task load), along with their overall sense of being pressured and/or threatened in some way with respect to what they are trying to accomplish (e.g., see Swain and Guttmann [Ref. 12] for a more detailed definition and discussion of stress and workload). To the extent crews or individuals expect to be under high workload, time pressure, and stress, this PSF is generally thought to have a negative impact on performance (particularly if the task is considered to be complex). However, the impact of these factors should be carefully considered in the context of the scenario and other PSFs that are thought to be relevant. For example, if the scenario is familiar, the procedures and training for the scenario are very good, and the rate at which the crew normally implements their procedures will allow them to achieve their goal on time, then analysts might decide that even relatively high expected levels of workload and stress will significantly impact performance. Although these factors may be difficult to measure, analysts should carefully evaluate their potential influence in the scenario being examined, before deciding on the strength of their effect.

8.   *Team/Crew Dynamics and Crew Characteristics [Degree of Independence Among Individuals, Operator Attitudes/Biases/Rules, Use of Status Checks, Approach for Implementing Procedures (e.g., Aggressive Crew vs. Slow/Methodical Crew)]*

This general factor can be extremely important, especially for in-control room actions where the early responses to an event occur and the overall strategy for dealing with the event develops. In particular, the way the procedures and conduct of operations manuals are written and what is (or is not) emphasized in training (which may relate to an organizational or administrative influence), can cause systematic and nearly homogeneous biases and attitudes in most or all crews, which can affect overall performance. Alternatively, ingrained crew differences in some areas can potentially lead crews to perform very differently under certain circumstances.

One of the most important lessons learned from performing ATHEANA HRA analyses in support of the PTS PRA studies [Ref. 13] was that the identification and understanding of crew characteristics is as important to appropriate HFE quantification as other more traditional elements of performance context (e.g., procedures, training). Aspects of how crews perform their jobs and interact with one another can create vulnerabilities in particular contexts. Examining crew characteristics generally involves evaluating the following factors (among others), which may be relevant to a specific context:

- how crew composition and organization (e.g., crew acting as team versus independent activities) can help or hinder appropriate response to the relevant context

- crew procedure implementation style or modus operandi (e.g., verbatim compliance; slow and methodical implementation; assignment of steps for independent performance; verification of step completion, either as performed, or at the end of performance; frequent use by crew members of accepted, contingent rules [i.e., continuous action statements in procedures]; and/or informal rules as shortcuts to steps that appear later in the procedure)

- communication habits (e.g., briefing strategies, ranging from EOP-driven to structured challenges called "BAGs" [the crew assesses what went on "Before," where they are "At" now, and where they believe they are "Going" in the scenario]) and protocols (e.g., formality of communications)

- how well informal rules help or hinder appropriate response to the relevant context

- crew preferences for key indicator screens (e.g., first-out panel, strip-chart displays, alarm screens)

- use/response of crews to input from operations staff (especially operator trainers)

- contributors such as "reluctance" or "hesitation" in implementing a particular mitigative strategy or action

The real key to gaining an understanding of crew characteristics is the integration of these various contributing factors, rather than their separate consideration. The resulting, holistic understanding is the necessary basis for realistically predicting operator behavior in specific accident scenarios. This understanding can represent a very different picture of expected crew behavior than would be produced from separate assessments of relevant training, procedures, and so forth. As a result of the PTS HRA studies, it became clear to the HRA team that understanding how operating crews approach the response to accident scenarios, how they think about the problem facing them, and how they work together are critical inputs to the HRA.

It is important to note that in performing the PTS analyses, it was found that there are wide variations in how procedures are implemented and used in the control room, and how problems are addressed. Common practices can vary across plant type (e.g., Combustion Engineering, Babcock & Wilcox, Westinghouse), across plants within a type, and across crews within a plant. Pacing of procedure implementation, allowable independent actions (within procedures), allowed aggressiveness in implementation (e.g., anticipating needed actions and jumping ahead when thought necessary), tendencies toward use of functional recovery procedures, guidelines for when it is acceptable to not follow procedures, confidence in procedures, and so forth, can all influence performance and be very different from plant to plant and, in some cases, from crew to crew within a given plant. Also, even the parameters that are emphasized for evaluating certain conditions can vary across plants and plant types. These findings emphasize the importance of the analyst's evaluation of crew characteristics and documentation of the findings in this area.

Furthermore, it is very important to evaluate how these identified crew characteristics might interact with the conditions in identified deviation scenarios later (in Step 6). In some instances, certain crew characteristics will protect crews from potential pitfalls created by somewhat unusual scenarios (e.g., the use of regular "BAGs," defined above). However, in other cases, certain characteristics may make the crews more susceptible, such as when mismatches occur between the scenario evolution and the way the crews implement their procedures. For example, successful response to a particular evolution may require crews to anticipate the need for an action and to begin preliminary preparation for that action before it is actually called out in the procedure.

It is critical that analysts observe simulator exercises, and conduct interviews and talk-throughs with operating crews and trainers, in order to obtain information relevant to this PSF. Obviously, unless the analyst is performing a relatively restricted PRA/HRA, it may not be possible to view simulator exercises for all scenarios. Nonetheless, observations of at least a few different types, along with crew debriefs and interviews with trainers, can be good sources of information. These activities can provide valuable insights into the overall crew response dynamics, styles, and attitudes.

A review of this factor should include asking the following questions (among others suggested above):

- Do the shift supervisors (SSs) differ in their leadership styles? Are some more democratic, and some more autocratic? Do they have the same kind of initial training, but is large variation accepted? Are there clearly stated goals as to how the SS should behave in this regard?

- Are the SSs trained to have an overview of the situation and call for crew meetings when needed? In the meetings, are the SSs told to always let the crew members speak first so that they are not unduly influenced by the SS, or do they take the lead and let the crew members confirm their assessment? What is the overall strategy for the meetings? Are the SSs taught to make decisions by themselves if there is no time for consultation?

- Are independent actions encouraged or discouraged among crew members? (Allowing independent actions may shorten response time but could allow inappropriate actions to go unnoticed until much later in the scenario.)

- Are periodic status checks performed (or not) by most crews so that everyone has a chance to "get on the same page" and allow for checking on what has been performed to ensure that the desired activities have taken place? In general, are there good communication strategies used to help ensure that everyone stays informed?

- In terms of communication protocol, are there any plant-specific guidelines for communication among the crew members? Are actions directed, and does the recipient of the direction then repeat the direction and report when the action is completed? Are all orders to be repeated and to contain an object and an action? Are all crews trained to communicate in this manner, but some appear to be uncomfortable with this level of formality and omit certain parts? (They might for example answer "yes" or "OK" instead of repeating the order or answering correctly to a question.) When crew members are asked to read a value, do they answer with the requested value and trend, even if the question could be answered with a "yes" or "no??

- Do most crews generally respond aggressively to the event, including taking allowed shortcuts through the procedural steps (which will shorten response times), or are typical responses slow and methodical ("we trust the procedures" type of attitude), thereby tending to slow down response times but reducing the likelihood of making mistakes?

- Are the crews attentive to continuous action statements in the procedures? Do the crews appear to anticipate upcoming events in the procedures that may require some initial early response?

- Are crews allowed to jump ahead in procedures or deviate from procedures if there is strong indication that they should?

In general, deciding whether the crew characteristics have a positive or negative effect will be contingent on the scenario being examined. For example, a particular style may be very positive for some scenarios, but not for others. Analysts should examine the various crew characteristics against the demands of the accident scenarios and HFEs/UAs being modeled, and strive to determine where crew characteristics may not fit well with the particular demands of the scenario. This evaluation will be relevant even for the assumed nominal case for the PRA accident sequence. In addition, analysts should also determine whether there is significant variability among the crews in terms of the characteristics above and evaluate whether any of those differences could lead to different results in responding to the given scenario? In other words, if crew characteristics vary, are vulnerabilities created for some scenarios? For example, some limited-time-frame scenarios may require relatively aggressive and anticipatory crews in order to be reliably successful. In instances such as this, during the HEP evaluation it is advisable to either assume the worst case for the crew differences in the scenario or the analysts may decide that it is important to model such aleatory influences separately. That is, probabilistically account for these aleatory differences more explicitly in the PRA model by modeling the HFEs/UAs under different EFCs. Guidance for more explicitly modeling such aleatory influences is provided later in this user's guide.

9. *Available Staffing/Resources*

For actions inside the control room, the available staffing and resources are generally not a significant consideration (i.e., not a particularly positive or negative factor) because plants are supposed to maintain an assigned minimum crew with the appropriate qualified staff available in or very near the control room.

However, for local actions outside the control room, this can be an important consideration, particularly depending on (1) the number and locations of the necessary actions, (2) the overall complexity of actions that must be taken, and (3) the time available and time required to perform the actions (see above for more on these related factors). For instance, where the actions are few, complexity is low, and available time is long, one or two available individuals may be more than enough to perform the required local actions and, thus, the available staffing can be considered to be adequate or even a positive factor. By contrast, where the number of actions and their complexity is high, and the available time is short, three or more staff may be necessary. Additionally, the time of the day the initiating event occurs may be a factor because night and "back" shifts typically have fewer people available than the day shift. As with crew characteristics (discussed above), if staffing levels differ significantly depending on the time of day, it is advisable to either treat the staffing level in an HEP evaluation as the minimum available depending on the shift, or probabilistically account for these aleatory differences more explicitly in the PRA model by modeling the HFEs/UAs under different EFCs. Guidance for more explicitly modeling such aleatory influences is provided later in this user's guide. It is incumbent on the analyst to determine whether the available staffing is sufficient to perform the desired actions and then assess the HEP(s) accordingly.

10.   *Ergonomic Quality of the Human-System Interface (HSI)*

The ergonomic quality of the HSI is generally not a significant factor relative to main control room actions because, given the many control room design reviews and improvements and the daily familiarity of the control room boards and layout, problematic HSIs have been remedied or are easily worked around by the operating crew. Of course, any HSI that is known to be very poor should be considered to be a negative influence for applicable actions even in the control room. For example, if common workarounds that are known to exist may negatively influence a desired action, the analyst should account for this influence in assessing important PSFs and eventually during quantification. Furthermore, it is possible that some unique situations may render certain HSIs less appropriate and, for such sequences, the analyst should examine the relevant interfaces.

However, because local actions may involve more varied (and not particularly human-factored) layouts and require operators to take actions in much less familiar surroundings and situations, any problematic HSIs can be an important negative factor for operator success. For instance, if reaching a valve to open it manually requires the operator to climb over pipes and turn the valve with a tool while laid out, or if in-field labeling of equipment is generally in poor condition and could lengthen the time required to locate the equipment, such non-ideal HSIs could become a negative PSF. Otherwise, if a review reveals no such problematic interfaces for the action(s) of interest, this influence can be considered to be adequate, or even positive if the interface somehow helps to ensure the appropriate response.

Walkdowns and field or simulator observations can be useful tools in discovering problems (if any exist) in the HSI for the action(s) of interest. Sometimes, discussions with the operators will reveal their own concerns about issues in this area.

11.   *Environment in Which the Action Needs To Be Performed*

Except for relatively rare situations, the environment in which the action needs to be performed is not particularly relevant to actions within the control room, given the habitability control of such rooms and the rare challenges to that habitability (e.g., control room fire, loss of control room ventilation, reduced lighting as a result of station blackout). However, for local actions, the environment could significantly influence operator performance. Radiation, lighting, temperature, humidity, noise level, smoke, toxic gas, weather (for outside activities, such as having to navigate a snow-covered roof to reach the atmospheric dump valve isolation valve), and other environmental factors can be varied and far less than ideal. Hence, the analyst should ensure that any HEP assessment includes a PSF to account for the influence of the environment in which the action needs to take place. This factor may be non-problematic (adequate) or a negative influence (even to the point of preventing the crew from performing the action). As with several other PSFs discussed in this user's guide (e.g., see crew characteristics and staffing), aspects of this PSF (e.g., weather effects on outside actions) may introduce aleatory influences that need to be addressed.

12. *Accessibility and Operability of the Equipment To Be Manipulated*

As with the environmental factor, the accessibility and operability of the equipment to be manipulated is generally not particularly relevant for actions within the control room, except in special circumstances such as loss of operability of controls as a result of a failure of the initiator or other equipment(e.g., loss of DC power). However, for local actions, accessibility and operability of the equipment may not always be ensured, and needs to be assessed in context with such influences as the environment, the need to use special equipment (discussed later), and the HSI. Hence, analysts should ensure that the HEP assessment includes a PSF to account for the accessibility and operability of the equipment to be manipulated. This factor may be non-problematic (adequate) or a negative influence (even to the point of preventing the crew from performing the required action).

13. *Need for Special Tools (Keys, Ladders, Hoses, Clothing Such as To Enter a Radiation Area)*

As for the environmental and accessibility factors, the need for special tools is not particularly relevant for actions within the control room, with the common exception of needing keys to manipulate certain control board switches or similar controls (e.g., explosive valves for standby liquid control injection in a BWR). However, for local actions, such needs may be more commonplace and essential to successfully perform the desired action. If such equipment is needed, the analyst should ensure that it is readily available, its location is readily known, and either it is easy to use or periodic training is provided, in order for this factor to be considered adequate. Otherwise, this factor should be considered to have a negative influence on operator performance, perhaps even to the point of making the failure of the desired action very likely.

14. *Communications Strategy and Coordination, and Whether One Can Be Easily Heard*

For actions in the control room, this factor is not particularly relevant, although the analyst should verify that the strategy for communicating in the control room is one that tends to ensure that directives are not easily misunderstood (e.g., the board operator is required to repeat the action to be performed, and then wait for confirmation before performing the action). In addition, the analyst should verify whether crew members avoid the use of double negatives. Generally, communication is not expected to be problematic; however, the analyst should account for any potential problems in this area (such as having to talk while wearing special air packs and masks in the control room during a minor fire).

For local actions, this factor may be much more significant because the environment or situation may be less than ideal. Thus, the analyst should ensure that the initiating event (e.g., loss of power, fire, seismic) or subsequent equipment faults are not likely to negatively affect the operators' ability to communicate as necessary to perform the desired action(s). For instance, having to set up the equipment and talk over significant background noise and possibly having to repeat oneself many times should be a consideration — even if only as a possible "time sink" for the action. Similarly, if "runners" are necessary to meet specific communication needs in the scenario of interest, the analyst should determine whether adequate staff will be available. Additionally, the operators should be adequately trained on the proper use of the communication equipment, its location should be readily known, and its operability should periodically be demonstrated. Depending on these characteristics, this factor may be non-problematic (adequate) or a negative influence (even to the point of preventing the crew from performing the required action).

15.   *Special Fitness Needs*

While typically not an issue for actions within the control room, special fitness needs could be a significant factor for a few local actions. Having to climb up or over equipment to reach a device, needing to move and connect hoses, or using an especially heavy or awkward tool are examples of where this factor could influence operator performance. In particular, the response time may be increased for successful performance of the action. In evaluating any HEP, analysts should consider activities that are physically demanding (or not), where it is appropriate to do so. Talk-throughs or field observations of the activities can also help to determine whether such issues are relevant to a particular HFE/UA.

16.   *Realistic Accident Sequence Diversions and Deviations (e.g., Extraneous Alarms, Outside Discussions, or Sequence Evolution Not Exactly Like That on Which Operators Are Trained)*

Particularly for actions within the control room, where early responses to an event occur and the overall strategy for dealing with the event develops, diversions and deviations can be an important factor. Through simulations, training, and the way the procedures are written, operators develop some sense of expectations as to how various sequences are likely to proceed, even to the extent of recognizing alarm and indication patterns and the actions that are likely to be appropriate. Differences between the actual sequence and how the scenario occurs in simulations — such as involving other unimportant or spurious alarms, the need for outside discussions with other staff or offsite personnel (such as fire fighters), differences in the timing of the failed events, behavior of critical parameters, and so forth — can all add to the potential diversions and distractions that may delay response timing or, in the extreme, even confuse the operators as to the appropriate actions to take.

Hence, analysts should examine the signature of the PRA accident sequence and the potential actions of interest by comparison to the operators' expectations, to determine if there is a considerable potential for such distractions and deviations. Observing simulations and talking with the operators can help in discovering such possibilities. This factor could impact the HEP mean value estimate, as well as the uncertainty in the HEP, which may be important in assessing the potential risk or establishing the limits for sensitivity studies with the HEP. Such considerations are the primary subject of the deviation analysis performed in Step 6 (described later) and, thus, are more directly handled during that step.

Use of Simulator Exercises to Support the Collection of Information for Task 2 of Step 5

An important activity to be performed to support the characterization of factors that could influence crew performance and lead to potential vulnerabilities is to observe simulator exercises for relevant scenarios (to the extent possible). This is highly recommended, where practical, for two general reasons:

(1)   HRA analysts can observe how operators behave and interact and hear how they think aloud about the task they are performing and what their approach should be (or obtain this information in debriefing following the exercise). In other words, analysts can observe and better understand (and thus account for) how elements of crew characteristics are integrated.

(2)   HRA analysts can observe operator responses to relevant scenarios and test any theories about operator response. In addition, general estimates of the timing of important diagnoses and actions (e.g., when particular steps in the procedure are reached) can be obtained and examples of how the operators understand and implement their procedures in particular scenarios can be observed.

As previously noted, the integration of crew characteristic factors is important. If simulator scenarios are carefully crafted, simulator exercises are among the best tools available to the HRA analyst for understanding crew characteristics and behavior, in general. We have found that simulator exercises can be useful at most stages of analysis, but a good understanding of the procedures relevant to the scenarios of interest facilitates selection of appropriate scenarios and observations and understanding during the simulations. Thus, it is suggested that at least an initial examination of PSFs and search for vulnerabilities should be done before setting up and conducting the simulator exercises. Collecting relevant information about crew characteristics will support the evaluation of PSFs against the nominal scenarios and facilitate the search for deviations that may cause crews problems (e.g., mismatches between their use of procedures and scenario evolution). Simulator exercises at the plant also can be used later in the analysis to validate or, at least, evaluate predicted crew behavior in some scenarios.

In addition, in the PTS HRA studies [Ref. 13] (as well as in an example application performed during the development of ATHEANA), there were some instances for which operator trainers from the plant were surprised by the response of their crews to a simulator scenario that was relevant to the HRA. Whatever the reason for such surprise, these instances emphasize the importance of performing scenario-specific simulator exercises to the extent possible, rather than relying upon expectations derived from procedures, training, and gross extrapolations from different simulator exercises.

Having said the above, it also is important to be aware of the limitations of simulator exercises. It has long been recognized that simulator exercises are not exact representations of "real world" events, so caution must be used in interpreting simulator results. Because the simulator cannot replicate all "real world" events, operator behavior in simulator exercises, similarly, should not be expected to exactly represent operator behavior in "real world" events. Consequently, it is recommended that observations of simulator exercises remain an input (albeit, a very strong input in some cases) in the overall HRA.

One example of the benefits of observing relevant simulator exercises occurred during the PTS PRA studies [Ref. 13], which revealed that because one plant crew was relatively slow and methodical in implementing their procedures (i.e., high trust in their procedures), the procedural steps relevant to addressing a somewhat unusual and fast-moving overcooling event were not reached until the potential for PTS effects was already present (at least some members of the crew were aware of these conditions fairly early in the scenario).

Summary of Output for Task 5.2

The output of Step 5, Task 2, is a summary or aggregation of the information collected. It is preferable that this information be systematically documented for use in later steps, especially Task 5.3 of this step. These characterizations will also be relevant with respect to the challenging contexts that will be developed starting in Step 6.

3.5.2.3 Task 5.3: Review Each PSF Against Each Accident Scenario and Relevant HFE/UAs To Identify Potential Vulnerabilities and Assess Their Potential Positive and Negative Influences on Performance in the Nominal Case

After collecting plant-specific information relevant to understanding the role of the PSFs and how they might bear on the scenario being examined, analysts perform the initial evaluation of the potential impact and importance of the various PSFs on the probability of the HFEs/UAs in the nominal scenarios for the PRA accident sequences. While decisions about the probabilities of the events are not made at this point, the goal is to determine which PSFs are likely to have a significant impact on the likelihood of success or failure, and why and how they would be expected to affect performance. For example, if the control room indications for the scenario are clear, the procedures are well-matched to the conditions, and the crews are regularly trained on similar scenarios, these factors would appear to be important and would positively support successful completion of the needed actions. However, even for the nominal case, other PSFs could negatively influence performance, and such factors need to be included as potentially important. For example, responding to the event may require relatively difficult actions outside the control room, and those actions would also have to be considered.

A list of the 16 PSFs to be considered is provided here to serve as a checklist for analysts:

(1)   applicability and suitability of training/experience

(2)   suitability of relevant procedures and administrative controls

(3)   operator action tendencies and informal rules

(4)   availability and clarity of instrumentation

(5)   time available and time required to complete the action, including the impact of concurrent and competing activities

(6)   complexity of the required diagnosis and response, the need for special sequencing, and the familiarity of the situation

(7)   workload, time pressure, and stress

(8)   team/crew dynamics and crew characteristics

(9)   available staffing/resources

(10)   ergonomic quality of the human-system interface

(11)   environment in which the action needs to be performed

(12)   accessibility and operability of the equipment to be manipulated

(13)   need for special tools

(14)   communications

(15)   special fitness needs

(16)   consideration of realistic accident sequence diversions and deviations (actually addressed in Step 6)

As discussed in the introduction to this step, in performing the HRA, analysts should also consider the potential for the above PSFs to create vulnerabilities that might become operative in deviations and related contexts from the nominal scenario, while evaluating their potential role in influencing operator failure or success for the nominal case. However, because the specific deviations may not yet have been identified (Task in Step 6), at least initially, analysts will need to make general judgments about aspects of these factors that might become relevant. For example, is the nominal scenario somewhat complex, such that it seems likely that if the scenario evolved a bit differently, the complexity might become an even greater problem. Analysts will have another chance to consider these PSFs during the deviation analysis in Step 6, after identifying plausible deviation scenarios.

These 16 PSFs comprise a set of influences that can be examined in the context of ATHEANA, used in investigating potential vulnerabilities and deviation scenarios, and eventually considered in quantifying HFEs and UAs in all scenarios. This is not to imply that all 16 of these PSFs will always be relevant or always contribute significantly to the likelihood of an operator failure, but rather that they have that potential. Usually, only a few will be major drivers of behavior, but which few are important will depend on the context for the specific human event being analyzed.

In addition, although an attempt is made to be reasonably complete with the use of these 16 PSFs, in thinking about influences on performance in the scenarios being examined, analysts (particularly in other domains) may identify other factors that they think could influence performance, or they may realize that certain combinations of factors may create unique influences. From the ATHEANA perspective, analysts should consider any factors they can identify (whether from the above list or not), in assessing crew performance. The goal is to identify factorsthat could potentially drive behavior, and then consider all of those factors in a holistic manner and subsequently estimating the probability of a UA or HFE during quantification. The expert opinion elicitation process used for quantification in ATHEANA (Step 8) does not impose constraints on which PSFs can be used.

Consideration of Aleatory Influences

In addition to evaluating the potential effects of the PSFs relative to the various nominal scenarios, in performing this task analysts should also investigate whether there are any potential aleatory influences associated with the PSFs, that they believe could have a strong impact on the likelihood of the HFEs/UAs. That is, to the extent aspects of the PSFs vary randomly, variations in the human action HEP could be expected depending on the specific context. For example, the analysts may determine that the differences in staffing levels in the plant between the day and night shifts could significantly affect the likelihood of a given HFE/UA, because of the number of personnel required to complete the action in the available time. Similarly, analysts may decide that it will require a reasonable level of aggressiveness on the part of crews to consistently respond within an appropriate time period for a given scenario, and they are aware that some plant crews may not always be adequately aggressive. That is, some crews tend to be very patient and methodical about applying their procedures, which in this particular scenario, may prevent them from consistently responding in a timely manner. When an event occurs and the particular type of crew on shift at the time would be functionally random and, therefore, their influence is considered aleatory.

Because the emphasis of Step 5 is on evaluating the expected effects of PSFs on the nominal case, the actual effect of the different "levels" of the aleatory influences will not be completely considered until Step 6. During the treatment of a given nominal scenario in this step (Step 5), when analysts identify a potentially important aleatory influence, they should pick the level or case of the aleatory factor that is either the most likely or the most expected to occur, and assume that level or case for the initial evaluation (e.g., assume the event occurs during the day shift). Other levels of the factor should be documented because they will be returned to later and considered for further analysis in Step 6. In Step 6, the emphasis is on identifying deviation scenarios with reasonable likelihoods of occurring (but probably with less likelihood than the probability of the conditions assumed for the nominal case) and that have a strong EFC. The deviation scenarios identified in Step 6 will also contribute to the aleatory uncertainty associated with the nominal scenario. Thus, treatment of the other levels of aleatory influences identified in this step will be completed in Step 6.

Dependencies Among Multiple HFEs/UAs in a Sequence

In evaluating the likely influences of PSFs in this step, it is also important to examine the effects of potential dependencies among multiple HFEs or UAs appearing in the same scenario. Successes or failures on the part of the operating crews earlier in a scenario can bear on the probability of success for later human actions. For example, if the crew has successfully performed all actions so far in the scenario, then there is reason to assume that they understand the scenario or are at least following the symptom-based procedures correctly. Alternatively, if they failed to perform a needed action or have taken an inappropriate action, this may indicate that they are confused about what is going on in the scenario. Thus, in evaluating influences on particular HFEs, analysts will also need to examine at least to some extent why crews may have failed on an earlier action (what could have lead them to make a given decision). Many of the PSFs discussed in this section could help explain why operators may take inappropriate actions (e.g., various biases, expectations, informal rules, etc.) and why they might persist on an incorrect interpretation of information. A more important analysis is whether significant additional indicators would be present that could support a correct decision for a given action, even if a human failure occurred earlier. The action associated with the original failure may itself be expected to produce feedback about plant conditions that could support a correct response later. New indications, changes in existing indications and trends, and new control room alarms should all be considered in evaluating the potential carryover effects from one event to another.

Clearly, information from plant operators and trainers can be used to evaluate the likelihood of success for an action given an earlier failure. Such influences will have to be considered in conjunction with the other PSFs and contextual information identified as being important in evaluating HFEs/UAs in a given nominal scenario.

3.5.3 *Output*

The output of Step 5 is a summary or aggregation of the information collected and evaluated for each of the HFEs/UAs associated with each of the nominal scenarios for the PRA accident sequences. In particular, the summary should include, for each HFE/UA being analyzed for each relevant scenario considering its nominal context:

- those PSFs and influencing factors that are particularly positive or negative with regard to their effect on the performance of the expected action

- a qualitative assessment as to the strengths of the above effects

- which factors seemingly will be the most important influences on the eventual HEP evaluation

- reasons for these judgments

- other observations about aleatory aspects of these factors that should be considered during the deviation analysis in Step 6

This information will be carried forward for use in the deviation analysis and for the derivation of HEPs in the quantification step. In addition, in Step 7, each HFE/UA in both the nominal scenarios and in any identified deviation scenarios will be evaluated for their potential to be recovered if the operating crew did take an inappropriate action. This potential for recovery will be considered during the quantification step in determining the final HEP for each HFE/UA. In fact, analysts may decide to complete Step 7 and Step 8 for the nominal cases and then return to the deviation analysis (Step 6) after that. Discussions during the quantification phase may facilitate the search for deviations. The output of this step should be systematically documented for use in the later steps.

> In this step, analysts are to take the following actions:
>
> - Consider the inherent time phases of the scenarios being addressed relative to the human action(s) of interest.
>
> - Collect or develop information relevant to determining how various PSFs might affect the performance of the action of interest for the scenario conditions.
>
> - Review each PSF to identify potential vulnerabilities to the human performance, noting those that are particularly important positive and negative factors to be considered in evaluating the action HEP.

3.6   Step 6: Search for Plausible Deviations of the PRA Scenario

3.6.1   *Purpose*

The purpose of Step 6 is to identify plausible deviations of each PRA scenario relevant to the HFE/UA being examined, and to carry forward the most important of these deviation scenarios and their associated EFCs to other steps of the ATHEANA methodology, so that the HFE/UA can be further analyzed considering these EFCs.

The search for deviations of each PRA scenario of interest and its nominal context (i.e., how the nominal scenario might evolve differently so as to be a greater challenge to the operators) is ATHEANA's most distinctive characteristic. Section 2 highlighted that when using ATHEANA, the analysts postulate other plausible scenario contexts, besides the defined nominal context that is associated with the scenario being considered in the PRA/HRA. These other plausible contexts focus on characteristics that may make the human error rate for the HFE/UA being analyzed higher than would otherwise be estimated for the human response to the nominal scenario.

ATHEANA, and particularly this step, is designed to address the premise that operator failures under the conditions associated with a modeled PRA scenario are more likely to result from deviations from the plant conditions generally expected by the operators for the PRA postulated accident scenario. Such expectations can be created by the procedures, the operators' training, and plant and industry experience. Deviations from what is generally expected, if sufficiently different, can cause serious mismatches between the actual situation and the operators expectations, their performance aids, their usual approach to implementing the procedures, and so forth. Step 6 is where the search for and development of possible deviation conditions that could be particularly problematic for the operators takes place.

3.6.2 *Guidance*

The analysts perform three tasks when implementing this step. The guidance is written for a single nominal scenario that is relevant to the HFE/UA being analyzed, and for which deviation scenarios are being postulated. The guidance is to be similarly applied to as many different nominal scenarios as are relevant to the HFE/UA of interest. These tasks are as follows:

(1)   Postulate plausible deviation scenarios.

(2)   Select that set of deviation scenarios that are judged to be the most risk significant for further analysis in subsequent steps in the methodology (i.e., screen out deviation scenarios not worthy of further analysis).

(3)   Document the implementation of Step 6.

The reader is encouraged to review section 2.3.2 in Section 2 of this guide for an overview of what this step is about and some examples of the types of differences that might be considered by the analysts when implementing Step 6. While we will explain the three tasks as separate entities, in practice and for efficiency reasons when performing Step 6, these are performed nearly simultaneously because of the strong linkages among the tasks. Further, and as part of Task 6.2, because Step 7 considers the potential for the relevant HFE/UA to be recovered before undesired plant consequences occur as a result of making an initial error, performing Step 7 while performing Task 2 of Step 6 is recommended. This will allow the analysts to screen out (i.e., not analyze further) deviation scenarios and their associated EFCs that can be easily recovered, making those scenarios risk insignificant. Refer to Section 3.6.2.2 for a discussion on all of the ways to screen out the deviation scenarios, of which consideration of recovery potential is only one.

3.6.2.1   Task 6.1: Postulate Plausible Deviation Scenarios

In this task, the analysts implement a formal search scheme designed to identify plausible deviation scenarios. These deviation scenarios are to represent how the situation could proceed somewhat differently from the nominal scenario such that the situation could be particularly troublesome for the operators, thus producing an error-prone (error-forcing) situation. This search scheme can be performed by the HRA analyst but he/she will likely need input from staff representing other disciplines (e.g., plant thermal-hydraulics analysts, operators and trainers, system engineers) in order to come up with the plausible deviation scenarios.

The information assembled from Step 5 and particularly any potential vulnerabilities identified during that step, serve as the primary input for Step 6. This is because Step 5 provides insights as to how operators may be vulnerable to mismatches between the potential error forcing conditions of a postulated deviation scenario, and the conditions for the nominal context for the PRA scenario. These insights allow the analysts to consider other credible but error forcing contexts that may cause such mismatches and therefore potentially make the human error rate much higher than the error rate for the corresponding nominal context.

The search scheme is designed to investigate two general categories of possible deviations:

(1)   postulated different plant conditions that might occur that could be caused by other random occurrences of equipment and human successes and failures, and yet still be represented by the modeled PRA scenario and, thus, impact the error rate for the HFE/UA

(2)   postulated crew/staff-related differences that may be present at the time of the scenario occurrence and, thus, impact the error rate for the HFE/UA.

The first set of postulations are made by the analysts using a series of guide words, similar to that used in HAZOPs (HAZard and OPerability studies) in the chemical processing industry, to examine "what if" situations. In this step, these guide words are used to investigate "what if" the (1) initiating event for the PRA scenario, or (2) the subsequent progression of the PRA scenario were to evolve differently than that defined for the nominal scenario. Put another way, when using the guide words, the analysts are asking themselves "how could the initiating event or the scenario progression be different in order to make the operator response associated with the HFE/UA being analyzed much more challenging and potentially error forcing?" The use of the guide words is enhanced by simultaneously considering (1) the vulnerabilities already identified for the nominal scenario in Step 5, as well as (2) how the PSFs discussed in Step 5 could become negative influencing factors (i.e., become vulnerabilities) if the plant conditions were somehow different. These two considerations, along with the guide words, are used to direct the identification of candidate deviation scenarios.

The second set of postulations involving crew/staff differences is performed by examining the PSFs discussed in Step 5 for which postulated differences would be manifested in terms of differences in the crew/staff. Differences in crew/staff characteristics may lead to the identification of additional deviation scenarios and contexts, as will the combination of both different plant conditions and different crew/staff characteristics.

Figure 3.6-1 illustrates the process being followed in performing this task of Step 6. More on how to implement this process is described next.

INPUTS ◄──────────────── PROCESS ──────────────►

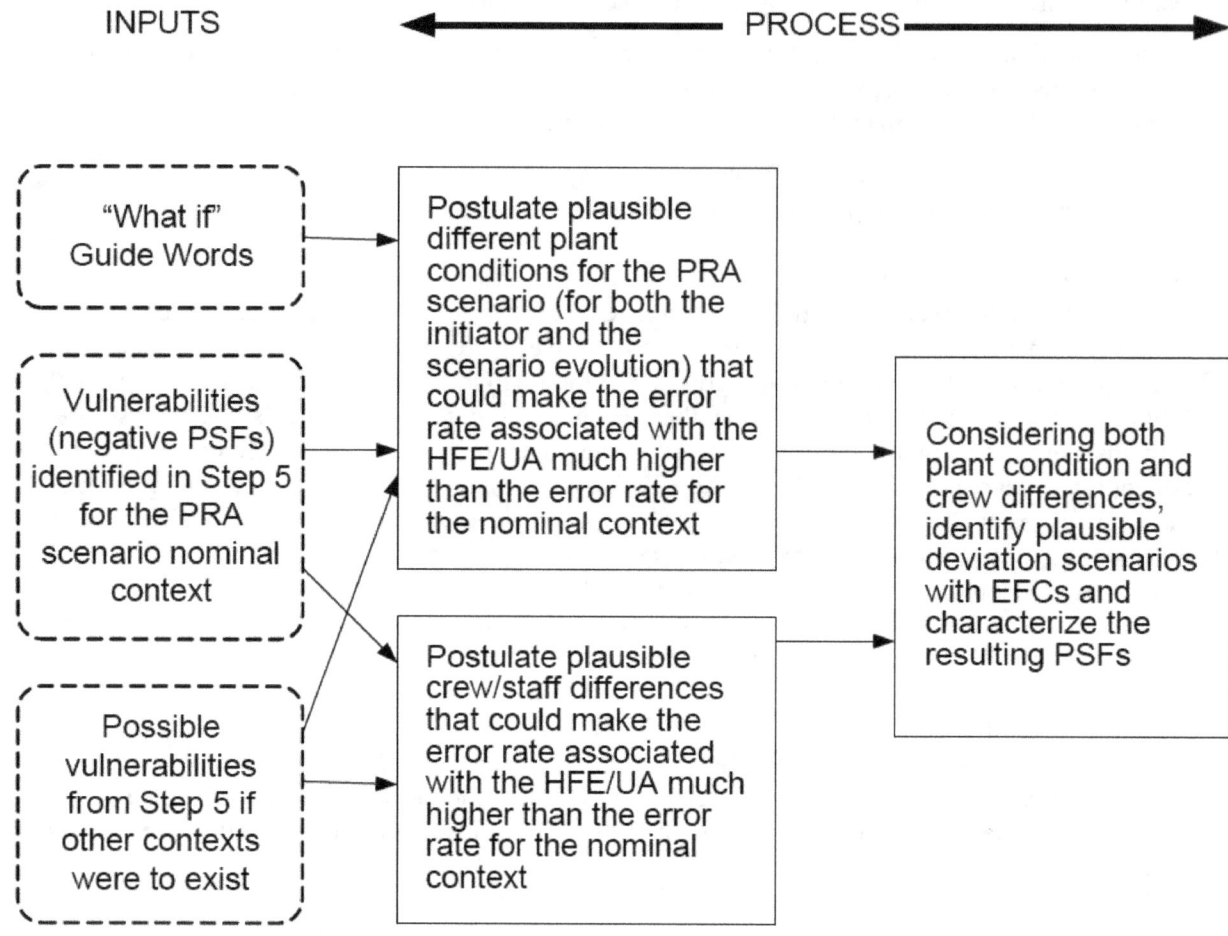

Figure 3.6-1. Implementation of Step 6, Task 1

The guide words used to conduct the first set of postulations (i.e., plausible different plant conditions) are provided below. Use of the guide words may provide overlapping suggestions for possible deviation conditions. That is acceptable, because such overlapping helps to ensure that a potentially significant set of conditions is not missed.

No or Not	A deviation in which something normally expected per the nominal context does not happen (e.g., what if the expected automatic low-level safety injection actuation did not occur?)
More or Greater or Larger	A deviation that represents a quantitative increase from that expected in the nominal context (e.g., what if the size of the breach were to be somewhat larger than that assumed in the nominal context?)
Less or Smaller	A deviation that represents a quantitative decrease from that expected in the nominal context (e.g., what if less flow than expected were available from the one operable train of injection such as if the train were operating in a degraded state?)
Early or Late or Never	A deviation that represents a change in the expected timing of events per the nominal context (e.g., what if the loss of injection occurred later in time, as a result of a room cooling fault, for example, rather than as a failure to start as assumed for the nominal context?)
Quicker or Slower	A deviation that represents a change in the expected speed or rate from that assumed in the nominal context (e.g., what if the vessel depressurization occurred much slower than that assumed for the nominal context?)
Shorter or Longer	A deviation that represents a change in the expected duration from that assumed in the nominal context (e.g., what if the battery power depleted in a shorter time than that assumed for the nominal context?)
Part of or Partial	A deviation in which only part of what is expected occurs (e.g., what if the stuck-open valve were only partially open rather than full open as assumed in the nominal context?)
In addition or As Well As	A deviation in which something additional occurs that is beyond what is assumed for the nominal context (e.g., what if other extraneous equipment faults and associated alarms were to also occur as well?)
Reversed	A deviation that is the logical opposite of that assumed for the nominal context (e.g., what if the stuck-open valve were to suddenly close on its own as a change to the nominal context?)
Repeated	A deviation that represents a repeated event (e.g., what if the relief valve was to open in a repeated fashion such as a second time during the scenario?)

The analysts apply these guide words to both the initiating event and to the scenario evolution in a thought exercise considering plausible changes from the nominal context for the PRA scenario of interest. That is, possible ways the initiating event for the PRA scenario of interest could occur differently, as well as ways the scenario itself could evolve somewhat differently, are investigated. To make the use of these guide words more efficient, they are applied by simultaneously considering the vulnerability information from Step 5, including ways all the PSFs (training, procedures, use of special tools, available cues/indications, environmental conditions, etc.) could become vulnerabilities in a different context. In this way, the use of the guide words is more efficient in discovering possible context differences that could be particularly troublesome for the operators because of the potential vulnerabilities.

For example, suppose it was found in Step 5 that the relevant procedure for the scenario of interest is written in a way that once a procedure step is implemented, there is little direction or other guidance to go back to that step and re-implement it if/when the corresponding condition related to that step changes [further, the crew might have exited from that procedure given the typical time to go through the procedure steps vs. the timing of the scenario]. Given this "vulnerability," the analysts could postulate a means by which the scenario could progress more slowly than assumed for the nominal context so that it could be more difficult for the crew to respond to the key parameter change given they have already progressed past the relevant procedure step or even exited the procedure.

In addition, Tables 3.6-1 and 3.6-2 provide some general characteristics that the analysts may find useful when considering how to postulate and consider different plant conditions during the application of the guide words. The set of scenario characteristics and their descriptions in Table 3.6-1 are based on those catalogued by Woods, Roth, Mumaw, and their colleagues (Refs. 14–18). They attempt to describe why scenarios with certain characteristics are difficult. The basic notion is that scenarios (which by definition evolve over time) contain features that create the opportunity for normal human information processing and action to be inappropriate or ineffective, essentially by creating unusual cognitive demands. Postulated deviation scenarios with such characteristics may therefore be particularly troublesome for the operators. Similarly, Table 3.6-2 provides additional characteristics and associated with questions to ask, but from a plant parameter perspective, that could cause troublesome conditions for the operators. Consideration of these characteristics could be beneficial when the analysts are attempting to postulate deviation scenarios with particularly error-forcing contexts.

Table 3.6-1. Scenario Characteristics that Can Cause Operators Problems in Detecting, Understanding, or Responding to a Situation

Scenario Characteristics	Description
Garden path problems	Conditions start out with the scenario appearing to be a simple problem (based on strong but incorrect evidence) and operators react accordingly. However, later correct symptoms appear, which the operators may not notice until it is too late.
Situations that change, requiring revised situation assessments	Once operators have developed a situation assessment and have started acting on it, it is often very difficult for them to recognize that there is new information or new conditions that requires them to change their situation assessment
Missing information	Key indicators may be missing as a result of failed sensors, lack of sensors, or lack of informants in the plant.
Misleading information	Misleading information may be provided as a result of inherent limitations of reports (e.g., stale information, inherent limitations of predictions, distortions resulting from indirect reports, secondary sources, translations).
Masking activities	Activities of other agents, or other automated systems may cover up or explain away key evidence.
Multiple lines of reasoning	Situations can occur where it is possible to think of significantly different explanations or response strategies, all of which seem valid at the time, but which may be in conflict (or a source of debate and disagreement by the operating crew).
Side effects	Situations can arise where the effects of human or automated system actions, or effects of the initial failure, have side effects that are not expected or understood.
Impasses	The scenario contains features where, at some point, it is very difficult for the operators to move forward, such as when procedures or the operators' situation model no longer matches the conditions, or assumed personnel or resources are not available.

Scenario Characteristics	Description
Late changes in the plan	The scenario is being managed according to a prepared plan, and then for some reason changes are required late in the scenario. Operators can become confused as to next steps; the plan is no longer well tested and can contain flaws, or the whole "big picture" gets lost by those managing the event.
Dilemmas	Ambiguity in the plan or in the situation (the event looks somewhat like two or more different accidents) can raise significant doubt in the operators' minds about the appropriate next steps.
Trade-offs	Operators must make impromptu judgments about choices between alternatives, such as when to wait to see if a problem develops (and may get out of control) versus jumping in early before it is clear what has caused the problem (just one of many examples).
Double binds	Conditions exist where operators are faced with two (or more) choices, all of which have undesirable elements.
High tempo, multiple tasks (Sub- or related categories are escalating events, cascading problems, and interacting problems)	The operators simply run out of resources (mental or physical) to keep up with the task demands. In escalating events, the problem keeps getting harder and harder or more complex. Cascading problems are those where the effects of one problem (or an attempt to solve it by the operators) create new problems. In interacting problems two or more faults interact to create complex symptoms that may have never been foreseen.
Need to shift focus of attention	As the scenario unfolds, the operators may need to move attention from one particular aspect of the problem to another, yet they remain focused on the initial problem area, which may be minor.

Table 3.6-2. Parameter Characteristics that Can Cause Operators Problems in Detecting, Understanding, or Responding to a Situation

Parameter Characteristics	Question
No indication	Does this scenario involve failed indicators? Does this scenario involve indications calculated from other failed instruments (e.g., subcooling based on RCS pressure)?
Small change in parameter	Within this scenario and with the existing human-machine interface design, is there a relevant parameter change small enough that it might be overlooked (i.e., not detected) such as a non-alarmed change in a valve position? Does this scenario involve small or significantly smaller-than-expected changes in any indication? Can the operators be led to a state of complacency by this small change? Within this scenario and with the existing human-machine interface design, is it likely that the operators will be misled by a small change as to the kind of situation they face (e.g., does it now resemble another scenario that is more familiar)? Does this scenario involve smaller-than-expected changes in an important parameter used as a cue or caution in the procedures, or used in training as a basis for actions? What is the likely effect of the operators misapplying this cue or caution? Can the operators be led to apply informal rules by this deviation? Can the operators be led to a state of complacency or forgetfulness by this small change?
Large change in parameter	Within this scenario and with the existing human-machine interface design, is there a relevant parameter change so large or out of range that it might be overlooked (e.g, indicator pegged at the top or bottom of a meter and not noticed). Does this scenario involve a large or significantly larger-than-expected changes in any indication? Can the operators be led to a state of anxiety by this large change? Within this scenario and with this interface design, is it likely that the operators will be misled by a large change as to the kind of situation they face (e.g., does it now resemble another scenario that is more familiar)? Does this scenario involve larger-than-expected changes in an important parameter used as a cue or caution in the procedures? Can the operators be led to apply informal rules by this deviation? Can the operators be led to a state of stress or anxiety by this large change?

Parameter Characteristics	Question
Lower or higher than expected value of parameter	Does this scenario involve indications that are lower or higher than would be expected? Does this deviation correspond with expected values for non-accident conditions, so that the deviation might not be detected as anomalous?
	Does this deviation correspond with expected values for other (different) accident conditions?
	Does this scenario involve lower or higher-than-expected values in an important parameter used as a cue or caution in the procedures?
	Can the operators be led to apply informal rules by this deviation?
	Can the operators be led to a state of complacency or forgetfulness by the lower change or a state of anxiety by the higher change?
Slow rate of change in parameter	Does this scenario involve slow or significantly slower-than-expected changes in any indication? Within this scenario and with the existing human–machine interface design, is it likely that the slow rate of change might be overlooked? Can the operators be led to a state of complacency or forgetfulness by this slow change?
	Within this scenario and with this interface design, is it likely that the operators will be misled by a slow change as to the kind of situation they face (e.g., does it now resemble another scenario that is more familiar)?
	Does this scenario involve slower-than-expected changes in an important parameter used as a cue or caution in the procedures? What is the likely effect of the operators mis-applying this cue or caution?
	Can the operators be led to apply informal rules by this slower deviation?
High rate of change in parameter	Does this scenario involve rapid changes in any parameter that, with the existing human-machine interface design, may be overlooked (e.g., fleeting changes, briefly appearing alarms or indications, or an indicator pegged at the top or bottom of a meter and not noticed)?
	Does this scenario involve rapid or significantly more rapid-than-expected changes in any indication? Can the operators be led to a state of anxiety by this rapid change?
	Does this scenario involve rapid changes in any parameter that, with this interface design, may be discounted or assumed to be anomalous (such as fleeting changes or briefly appearing alarms or indications)? If overlooked or ignored, is the absence likely to confuse the operators as to the kind of situation they face (e.g., does it now resemble another scenario that is more familiar)?
	Does this scenario involve faster-than-expected changes in an important parameter used as a cue or caution in the procedures?
	Can the operators be led to apply informal rules by this deviation?

Parameter Characteristics	Question
Changes in two or more parameters in a short time	Does this scenario involve changes in two or more indications that are significantly different from expected? Do they involve rapid changes in any parameters that, with this interface design, may be overlooked (such as fleeting changes or briefly appearing alarms or indications)?
	Does this scenario involve changes in two or more indications that are significantly different from expected or inconsistent? If observed, will these indications cause operators to be significantly uncertain or confused as to the situation in the plant?
	Does this scenario involve rapid changes in any parameters that, with this interface design, may be overlooked (such as fleeting changes or briefly appearing alarms or indications)? If overlooked, is their absence likely to confuse the operators as to the kind of situation they face (e.g., does it now resemble another scenario that is more familiar)?
	Does this scenario involve changes in two or more indications that are significantly different from the procedural expectations? If observed, will these indications cause operators to be significantly uncertain or confused as to how the procedures should be applied to the plant?
Delays in changes in two or more parameters	Does this scenario involve changes in two or more indications that are significantly delayed from what is expected? Do they involve late changes in parameters that, with this interface design, may be overlooked?
	Does this scenario involve two or more indications that are significantly delayed from what is expected? If observed, will these delayed indications cause operators to be significantly uncertain or confused as to the situation in the plant?
	Does this scenario involve changes in two or more indications that are significantly delayed from what is expected?
	Do they involve late changes in parameters that, with this interface design, may be overlooked? If overlooked, is their absence likely to confuse the operators as to the kind of situation they face (e.g., does it now resemble another scenario that is more familiar)? Delayed information can be ignored or reinterpreted to match earlier (premature) assessments of the plant situation (such as being dismissed as "instrument error").
	Does this scenario involve significant delays in two or more indications compared with the procedural expectations? Will these delays cause operators to be significantly uncertain or confused as to how the procedures should be applied to the plant?

Parameter Characteristics	Question
One or more false indications	Does this scenario involve false indications that, together with the genuine indications, resemble a situation that is expected (i.e., consistent with other on-going activities that could lead operators to ignore or not attend carefully to the indications)? Does this scenario involve false indications that, together with the genuine indications, resemble a situation that is expected (i.e., consistent with other on-going plant activities that could explain their presence)? Will these false indications cause operators to be significantly uncertain or confused as to the situation in the plant? Does this scenario involve false indications that mislead the operators into believing that the required actions are no longer necessary or are not possible (e.g., false indication of a caution or prohibition)? Does this scenario involve false indications that require inconsistent actions by operators (e.g., both depressurize and repressurize the primary system)?
Direction of change in parameter(s) over time is not what would be expected (if the nominal scenario was operative vs. the deviant) Direction of change in parameters over time, relative to each other, is not what would be expected (if the nominal scenario was operative vs. the deviant) Relative rate of change in two or more parameters is not what would be expected (if the nominal scenario was operative vs. the deviant)	Does this scenario involve changes in one or more parameters over time that are significantly different than what would be expected if the nominal scenario was operative as opposed to the existing deviant scenario. If observed, will these changes cause operators to be significantly uncertain or confused as to the situation in the plant?
Behavior of apparently relevant parameters is actually irrelevant and misleading	Does this scenario involve the occurrence of one or more parameters that are actually irrelevant and misleading given the deviant scenario being examined. If observed, could these parameters cause operators to be significantly mislead. Would they be similar to patterns that would occur in the nominal scenario.
Parameters indicate response for which insufficient resources are available or indicate more than one response option	Does this scenario involve a situation where the unavailability of resources make the response difficult to execute? Are there competing options or options with trade-offs?

In the second set of postulations covering crew differences, rather than focusing on plant condition differences, the thought exercise considers differences among crew/staff characteristics that could be important to the human response for the scenario of interest.

Some of the PSFs that may be particularly relevant for considering crew differences include the following examples:

- differences in some tendencies and informal rules among the different crews

- differences in the crew communication protocols

- differences in such crew characteristics as degree of independence allowed among the operators and preferences regarding the use of indications and computer screens

- differences in the strategies used among the crews as to how methodical the procedures are used and how often crew-wide checks of plant status are or are not used

- differences in the staffing levels for the different shifts, etc.

Because it is not known when the initiating event will occur, which crew is responding to the event could make a significant difference in the human response.

When doing the above thought exercises covering both postulations, the analysts should consider both the diagnosis and execution portions of the HFE/UA being examined because some deviations may primarily affect the diagnosis portion of the HFE/UA, others may affect the execution portion, and still others may affect both diagnosis and execution. For example, possible changes in the available cues and indications might typically affect the ability of the operators to diagnose the need to take the desired action if, for example, a key indication were to fail or be otherwise unavailable as a result of testing or calibrating, or if it were involved in a typical workaround. Instead, a possible change regarding the availability of a special tool (suppose it is not where it should be) might significantly affect the ability and timing of an operator to perform an ex-control action requiring the use of that tool.

Task 6.3, discussed later, covers how to document the implementation of Step 6 and provides an illustration of documenting such a thought exercise.

3.6.2.2 Task 6.2: Screen Out Possible Deviations Not Worthy of Further Analysis

The above process may produce an unmanageable number of possible deviations that might be particularly troublesome for the human action of interest. Thus, it is important that while performing Task 6.1, that Task 6.2 be conducted somewhat simultaneously so that screening out of less important deviations is part of the ongoing thought exercise. In this way, only those plausible deviations involving the most error-forcing contexts are carried forth in the ATHEANA methodology. This is important because the number of HFEs/UAs that have to have their HEPs estimated is a function of how many EFCs survive this task. For instance, even if every EFC has the same UA and only one UA is associated with each context, but there are 16 EFCs, then the UA will have to quantified 16 different times to cover the 16 EFCs (plus one more time to cover the nominal context directly from Step 5). So it is incumbent on the analysts to be very selective and carry forth only those plausible deviations and their associated EFCs that have the most promise in contributing to high HEPs (because the contexts are very error-forcing).

The following screening criteria should be used in selecting the most important deviation scenarios to carry forward:

- Is the perceived strength of the combined negative PSFs for the postulated deviation scenario very high, such that the context is potentially among the most error-forcing of those considered?

- Is the recovery potential from Step 7 (best done in parallel with this task) judged to be low so that if the initial error represented by the HFE/UA were to be made, it does not seem likely that the operator(s) would recover from the mistake before undesired consequences occur?

- Is the likelihood of the postulated deviation scenario and its associated EFC sufficiently high that it is worthy of being carried forward in the analysis rather than being so low that even with a corresponding high HEP, the overall contribution to risk will be insignificant?

- Are their similarities among the postulated deviation scenarios and their associated EFCs so that some can be combined thereby lessening the number of contexts to be addressed?

The first screening consideration examines what PSFs might be negative for a postulated context and the strength of those negative effects. The more PSFs simultaneously affected and/or the stronger the negative influences, the more the postulated deviation scenario and its associated EFC is worthy of having the HFE/UA be assessed (i.e., the HEP estimated) for that EFC.

As discussed in Step 7 (see that step), the second screening consideration addresses the likelihood of recovering from an initial error. If the recovery potential is judged to be quite low for the EFC, the more the postulated deviation scenario and its associated EFC is worthy of having the HFE/UA be assessed (i.e., the HEP estimated) for that EFC.

The third screening consideration considers the likelihood of the deviation scenario and its EFC. For instance, if the postulated deviation requires the unavailability of a certain indicator during the scenario, and that the most methodical/slowest crew be on shift that would likely take the longest to get to a particular procedure step, and that an additional specific equipment failure also needs to occur during the course of the evolution of the scenario, it may be that such a combination of events is too unlikely. In such a case, even if the corresponding HEP for the HFE/UA were high (say >0.1), the risk significance of this HEP would be low because of the very low probability that the PRA scenario involves these other conditions.

Finally, the fourth screening consideration is very similar to that commonly done when extending the results of a Level 1 PRA to a Level 2 PRA. Just as core damage sequences with similar characteristics are combined to limit how many are addressed in the Level 2 portion of the analysis, the analysts should look for EFCs that have similar enough characteristics in terms of possible impact on the operator performance. In such cases, these should be combined so that the number of deviation scenarios and their associated EFCs is distilled down to a manageable set.

The overall goals are (1) to be very selective regarding how many and which deviations are analyzed further for the HFE/UA of interest, but (2) not screening out a unique and potentially very risk-important deviation/EFC.

3.6.2.3 Task 6.3: Document the Performance of Step 6

Step 6 is a process of postulating differences from the PRA-relevant nominal context, both for differences in plant conditions as well as differences in crew/staff characteristics. It also involves screening out relatively unimportant differences that might be postulated. To record these thinking exercises, it is important that the analysts document these postulated differences, and why certain ones are screened out from further analysis.

In this step, the analysts should perform the following tasks:

- Postulate plausible deviation scenarios that are judged to have sufficiently strong EFCs.

- Screen and/or combine these scenarios so that the potentially most risk-significant of these are kept for further analysis.

- Document the implementation of the above process.

Table 3.6-3 provides an illustration of how analysts might document the postulations discussed above. Once plausible deviations are identified, the documentation should also summarize which deviation scenarios were ultimately selected and/or combined for further analysis in the remaining steps of the ATHEANA methodology.

3.6.3 Output

The output of Step 6 consists of the deviation scenarios that will be carried forward in the analysis for which the HFE/UAs will have their HEPs estimated for the EFCs that are associated with the selected deviation scenarios. In particular, the following should exist:

- sufficient documentation covering the deviation scenarios considered and the resolution of those scenarios (e.g., which were screened out and why; which were combined together and why)

- for those scenarios selected to be carried forward in the analysis, a description of the deviation scenario and its EFC noting what is particularly different from the PRA scenario nominal context

- for those scenarios carried forward in the analysis, a summary of the potential significance of the deviation scenario in terms of the potential impact on the operator performance. This should include what PSFs are judged to be most negatively affecting the human response associated with the HFE/UA being addressed, given the deviation scenario and its EFC

Table 3.6-3. Deviation Scenario Considerations

PRA-Related Nominal Context

Loss of condenser reactor-turbine trip with MSIV closure followed by an immediate stuck-open atmospheric dump valve leading to a potential severe cooldown event. Operators successfully isolate feed to the affected steam generator in this sequence.

HFE of Concern

Failure to close the corresponding manual isolation valve for the stuck-open atmospheric dump valve within 30 minutes. This requires a local action, on the roof of the turbine building where the area is unprotected from the elements, using a reach-rod through a concrete barrier for personnel protection from the atmospheric dump valve release.

Possible Plant Condition Deviations

Guide Word: In Addition or As Well As

Possible Deviation: In addition to the sequence occurring as described, it happens while there is snow/ice/bad weather conditions on the turbine building roof where this action takes place. For the nominal context, such an adverse condition was not assumed.

Potential Significance of Deviation

This additional adverse condition could delay the action because of the local operator tripping, falling, or otherwise having to take time to take precautions when preparing to carry out the activity on the roof. This deviation has the combined dependent effect of making the environment PSF and, hence, the Time Available vs Time to Act PSF becomes more negative than assumed for the nominal context. The likelihood of snow/ice on the roof is not that infrequent (~25% of the year). No relevant recovery actions noted.

Possible Crew/Staff- Related Deviations

Possible Deviation

Considering the range of crew speed and the differences observed (during simulator training) as to their willingness to anticipate actions rather than wait until they get to the procedure step, some control room crews have shown a tendency to anticipate the need to contact a local operator while still in EOP E-0 (reactor trip) while other control room crews typically wait until they get to the appropriate step in EOP E-6 (excessive cooldown). Which crew is on shift at the time of the scenario could affect the HEP for this HFE. For the nominal context, the anticipatory style was assumed as this is typical of most of the crews.

Potential Significance of Deviation

This deviation is associated with the Crew Characteristics PSF and in particular, the anticipatory vs. methodical style of the various crews. The crews with the more anticipatory style of response, will allow more time for the local operator to travel to the turbine building roof and execute the action. For the crews with the more methodical, procedure-following style of response, the Crew Characteristics PSF is more negative for this scenario in that it will allow far less time (perhaps as much as 10–15 minutes of the total 30 minutes available) if the local operator is not contacted until EOP E-6 has been entered. This has a direct effect on providing little margin for the local operator to perform the isolation within the required 30 minutes, especially if he/she has not been notified to do so until 10–15 minutes have already gone by (i.e., the time available vs. time to act has much less margin). Approximately 20% of the crews tend to follow this more methodical approach. No relevant recovery actions noted.

Table 3.6-3. Illustration of Deviation Scenario Considerations (cont'd)

Any Deviations Worthy of Further Analysis?

Postulate a single deviation scenario with both of the above deviations, together, as additional characteristics to the nominal scenario.

It is noted that both of the above postulated deviations represent plausible differences from that assumed for the nominal context in that the likelihoods of both conditions are not small [i.e., the combined deviation context frequency = nominal context frequency x 0.25 (snow/ice) x 0.2 (methodical crew style)]. Further, both affect the margin available between the time available to take the desired action (30 minutes) and the time it takes to travel to the roof and perform the action. The environment deviation (snow/ice) could slow down the execution time because of the greater difficulty in carrying out the isolation and the extra care needed to avoid injury to the local operator. The crew characteristic deviation could delay the time when the local operator is notified to take the desired action, thus lessening his/her time available to actually execute the action. The HEP for this HFE (particularly the execution portion of the HFE) could thus be significantly affected as compared to that for the nominal context.

3.7   Step 7: Evaluate the Potential for Recovery

3.7.1   *Purpose*

The purpose of this step is to address the recovery potential for the HFE or UA being analyzed in the context of each nominal scenario documented in Steps 3–5 and each credible deviation scenario that has been postulated and selected for further analysis in Step 6. This step addresses the HLR-HR-H high-level requirement in the ASME PRA Standard [Ref. 2].

> In Step 7, the analysts examine the potential for recovery of HFEs/UAs in both the nominal and deviation scenarios. Deviation scenarios might be dropped from further analysis if there is a high likelihood of recovery.

For a given deviation scenario, if the likelihood of the operators recovering from their initial error (without undesired changes in plant conditions occurring) is considered to be very high (a qualitative judgment by the analysts), then the deviation scenario, the associated context (EFC), and the resulting HFE/UA is much less important. That is, if any initial error can be qualitatively justified by the analysts to have a very high likelihood of being recovered before undesired plant consequences occur, then the risk of the initial error is averted. In such cases, the risk contribution of the HFE/UA, when considered with recovery, should be small and can therefore be dropped from having to undergo detailed quantification analysis (i.e., estimating the HEP for the HFE/UA with recovery included would not be necessary). Two exceptions to this guideline would be that 1) the likelihood of the deviation scenario and its EFC is relatively high (i.e., close to or higher than that for the nominal scenario), so that the risk contribution of the HFE/UA for the deviation scenario and its EFC could still be relatively significant, or 2) the issue or HRA application requires a quantitative estimate (or it is desirable to do so) even though there is a high recovery potential.

For the HFEs/UAs being addressed for the nominal scenarios, analysts will always carry the HFEs/UAs forward to quantification and the information obtained here will be used during the quantification process. That is, the potential for recovery, along with the context of the events identified in Step 3.5 (important PSFs, across scenario dependencies, etc.) will all be considered together in obtaining the final HEP for the HFEs/UAs. This is done so that a base estimate of the risk contribution of the HFE/UA is quantified, including the potential for recovery, for the nominal scenario.

The above treatment of HFEs/UAs for nominal and deviation scenarios is illustrated conceptually in Figure 3.7-1.

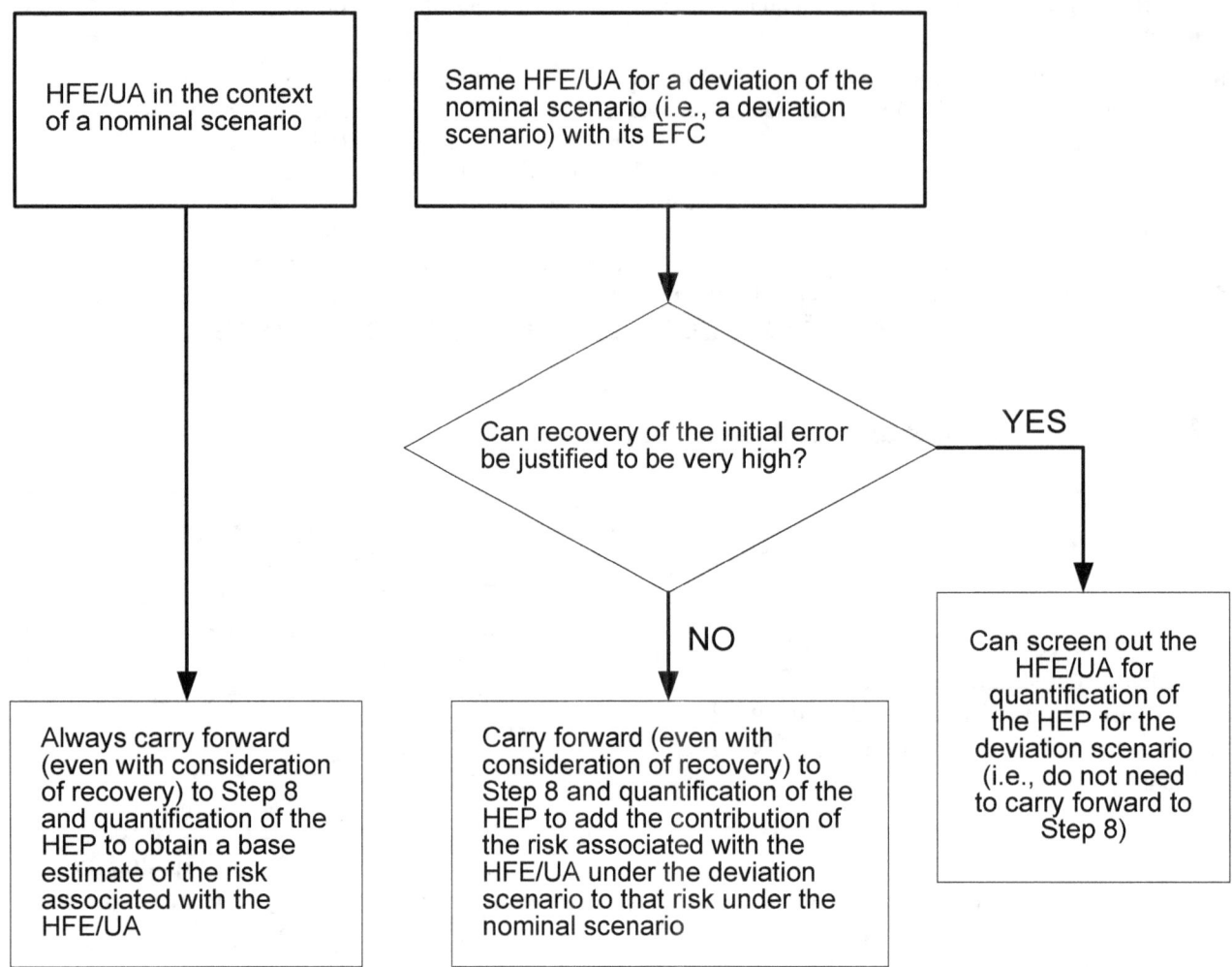

Figure 3.7-1. Concept of When HFE/UA is Carried Forward for Further Analysis in Step 8 (Quantification of the HFE)

3.7.2 *Guidance*

This step is carried out by performing the following four evaluations for each HFE/UA and associated scenario. As in Steps 5 and 6, analysts must keep in mind the entire context (important PSFs, plant conditions, dependencies) identified for the scenario, in assessing the potential for recovery of an HFE/UA, whether an EOO or an EOC:

(1)   Define the possible recovery action(s) given the initial error corresponding to the HFE/UA has occurred.

(2)   Consider the time available to diagnose the need for and perform the recovery action so as to avoid a serious or otherwise undesired condition.

(3)   Identify the existence and timing of cues as well as how compelling the cues are that would alert the operators to the need to recover and provide sufficient information to identify the most applicable recovery action(s).

(4)   Identify the existence and timing of additional resources (e.g., additional staff, special tools), if necessary, to perform the recovery.

Thus, the description of each scenario is extended using the information obtained in the evaluations described above, in order to justify the judgment of either a high or low recovery potential. In assessing the likelihood of recovery (both now qualitatively and during quantification if carried forward to Step 8), the analysts should be careful of being overly optimistic and should keep in mind the following:

•   There may be dependencies between the initial error and the identified recovery that could make the likelihood of recovery low (i.e., the failure to recover is high).

•   Initial mindsets of the situation are sometimes hard to break, particularly if there was a strong EFC.

•   A pattern of incorrect actions across the scenario could certainly weigh against the likelihood of recovery.

•   Distractions or attention to other activities could cause new cues to be missed or overlooked.

•   The operators may delay taking a recovery action because there are negative consequences with taking the action (e.g., a release of reactor coolant into the containment), especially if other plant hardware that provides an alternative action is "almost available" or "almost fixed."

With the above information in mind, for the deviation scenarios, analysts can make assessments of the likelihood of recovery given the EFC. This does not have to be a specific quantified value (unless it is desired to do so), but can be a qualitative assessment such as "very high likelihood of recovery." This information may then be used in conjunction with an estimated likelihood of the deviation scenario and its EFC, to support a decision to drop the HFE/UA evaluations for the deviation scenarios from further consideration or to carry them forward for quantification. It is best to document this screening process for future uses of the HRA.

For the nominal scenarios, the obtained set of information and initial judgments about the likelihood of recovery of a given HFE/UA will be carried forward for quantification.

3.7.3 *Output*

The output of Step 7 includes expansions and finalization of nominal and deviation scenarios and their descriptions, as necessary, to incorporate that information needed to understand what recoveries were considered and the recovery likelihood assessment. Some deviation scenarios and the associated HFE/UA may be dropped from quantification if the recovery potential is very high. Sufficient documentation should be created that provides justification for dropping such scenarios from further analysis.

*   *   *   *   *   *

This completes the qualitative portion of the ATHEANA HRA method. In performing Steps 1–7, considerable qualitative insights should have been gained regarding the HFEs or UAs being addressed including not only the typical EOOs addressed in PRA/HRA, but also possible EOCs. The investigative process of considering other relevant and credible contexts (not just the nominal context for the PRA scenario) allows one to better understand under what conditions the human error of interest might be made and what may drive the error rate depending on the specific context associated with the situation.

With this knowledge about the HFE or UA of interest, quantification must be able to account for all this knowledge. Hence, since the publication of NUREG-1624, Rev. 1, a quantification technique has been developed that uses an expert elicitation process that can take advantage of the entire knowledge base gained in performing the above seven steps. Guidance on performing the quantification is addressed in the next step, Step 8.

3.8   <u>Step 8: Estimate the HEPs for the HFEs/UAs</u>

3.8.1 *Purpose*

Quantification as part of HRA involves the derivation of HEPs (i.e., probability distributions for the error rates assigned to the HFEs). The purpose of this step is to estimate the HEPs for the corresponding HFEs/UAs being analyzed, for each of the contexts (i.e., the nominal context and any EFCs) carried forward from previous steps in ATHEANA.

The reader is referred to Section 2.3 before implementing Step 8. That section has already provided an overview of key distinctions of ATHEANA with regard to possibly breaking down HFEs into UAs (including perhaps, EOCs) in Section 2.3.1, addressing multiple contexts in Section 2.3.2, and in particular, providing the mathematical formulation of how the different contexts, UAs, and corresponding HEPs are put together in Section 2.3.3. As a reminder, the most general form of the mathematical formula is repeated below:

$$P(HFE \mid S) = \sum_{j} \sum_{i(j)} P(EFC_i \mid S) \times P(UA_j \mid EFC_i, S)$$

In this step, the individual HEP is being estimated for each UA for each given context (shown here as the term $P(UA_j|EFC_i,S)$). A simple example of the use of this formulation was also provided in Section 2.3.3. The HRA/PRA analysts should be familiar with the information provided in Section 2.3. In this step, estimates of the necessary HEPs are made so as to supply the error rates that are needed to produce quantitative risk evaluations.

Step 8 addresses the HLR-HR-G requirement as well as the quantitative treatment of including recovery actions (see Step 7) per the HLR-HR-H high-level requirement in the ASME PRA Standard [Ref. 2].

3.8.2   *Quantification Approach and Related Technical Issues*

In ATHEANA, each UA (including any EOCs) associated with an HFE is assessed for each context being carried forward in the analysis. This is done using a structured expert elicitation process. In estimating each HEP, the experts consider the plant conditions and relevant PSFs (including whether the PSFs are positive or negative influencing factors as well as their relative strengths) associated with each context in a holistic and integrated manner, and ultimately arrive at an estimate for each HEP.

Because the current approach to estimating the HEPs for the HFEs/UAs was not addressed in the original NUREG-1624 [Ref. 1], it is appropriate to discuss the quantification approach in some detail including who should be involved, controlling for biases when performing elicitation, addressing uncertainty (much of the theoretical discussion on this topic is not unique to ATHEANA but is germane regardless of the HRA method being used), and other issues associated with eliciting probabilities, before providing the guidance for implementing Step 8.

The approach for quantification in ATHEANA relies on a very structured, facilitator led, expert opinion elicitation where experts provide their review and insights on the factors judged (from previous ATHEANA steps) to be driving performance, along with their judgments as to the appropriate estimate for the HEP. While proper qualifications of the experts, as the appropriate team, has been generally addressed in the introductory part of Section 3 of this document, it is important to note several key roles in the elicitation phase of the ATHEANA process.

3.8.2.1   Who Should the Experts Be?

The best expertise comes from plant people with first hand knowledge that is pertinent to the elicitation. If at all possible, operators must participate. They bring, perhaps, the most appropriate knowledge about the plant and how it behaves, how they interface with the plant equipment, operations protocols that may be relevant, their level of familiarity with the use of the procedures under a variety of circumstances, the applicability of their training as different contexts are explored, and so forth. It is important during the elicitation that they see the issue in terms similar to what they would see in the plant, rather than only in the context of a PRA scenario leading to core damage.[9] Additional expertise comes from trainers, procedure writers, engineering analysts, and PRA experts.

---

[9]   If you ask an operator, "What is the chance that you might fail to start a needed pump?" the operator will tell you why he/she cannot possibly fail to start it. If you create a situation/scenario up to this point, give the operator a set of indications and ask "What will you do now?", the answer may be quite different.

In particular, in the applications of ATHEANA to date, it has been observed that an especially important point of view is that of the plant's operator trainers, who can reflect on the wide range of operator characteristics and the experience of observing many crews in many situations. They can help, for instance, identify what conditions can cause operators to lock onto an incorrect situation assessment or be misled by unexpected indications.

Finally, the integrity of the process depends on one member of the team we will call the facilitator. This person must ensure that the caveats and controls mentioned in the remainder of this section are observed. The role of the facilitator is described in a text box below; it is a role one of the PRA team might fulfill. It requires a little study of the process and an understanding of some of the issues that can lead to improper judgments by the experts. The facilitator needs to thoroughly study the biases described in the text box "Controlling for Unintentional Bias," and be on guard to prevent them from corrupting the elicitation. The detailed elicitation steps under the guidance for Step 8 include many helpful controls that work well, as long as the facilitator follows them.

As a practical matter, it is recommended that the number of experts be at least 3 persons, and perhaps up to 6 in number. The goal is to represent a sufficient breadth of experience, but without having so many people as to unduly detract from carrying out the process or make it take too long.

3.8.2.2 Addressing Uncertainty

There are many sources of uncertainty associated with human performance in power plant operations. ATHEANA, by identifying and separating out each error-forcing context along with the nominal context, accounts for a significant portion of the uncertainty (particularly the aleatory uncertainty) in HRA quantification explicitly. Nevertheless, no matter how finely the analysts may distinguish contexts, some residual uncertainty remains, both aleatory and epistemic. Failing to acknowledge uncertainty and account for it can make any evaluation suspect, particularly one that relies heavily on expert elicitation. Therefore, it is important to be explicit about what is being addressed and to incorporate a careful treatment of uncertainty in the elicitation process.

The Facilitator

A facilitator is a normative expert with the interpersonal skills to control the elicitation process and ensure that it puts all available information on the table, and that the experts are fairly heard and not allowed to hide behind others.

By understanding how inadequacies in probability estimation and biases occur (see text box on bias), the information can be used to combat their influence. The inadequacies of individuals can be dealt with by selecting analysts with a variety of expertise and by facilitating the process, including challenging participants to explain the basis for their judgments. A facilitator can directly address biases. For example, representativeness bias involves ignoring available information and replacing a careful evaluation of that information with quick conclusions based on too much focus on part of the information or allowing irrelevant information to affect conclusions. The facilitator must challenge analysts, asking them to explain their opinions. The facilitator must use his own judgment to sense when an individual is not using the full information.

Moreover, by understanding the heuristics that people often use to develop subjective probability distributions and the biases that attend those techniques, that awareness can help experts and analysts avoid the same traps. Through understanding which framings for eliciting distributions cause problems, we can use those that work better. Because the facilitator is familiar with the potential biases, he/she can test the group's ideas and push them in the right direction. The strategies presented below should be used either explicitly or implicitly through the questioning of the facilitator, as described in the SSHAC report [Ref. 19]. In addition, Tversky and Kahneman [Ref. 20] give many detailed examples useful for helping facilitators develop awareness of such useful aids. Some of the simplest and best aids include:

- constructing simple models of the maximum and minimum points of a distribution, avoiding focus on the central tendency until the end points are studied to avoid anchoring (test these models to examine the evidence supporting them rather than relying on opinion alone)

- seeking consensus on the evidence considered by the experts

- testing distributions by asking if the assessor agrees it is equally likely for the real answer to lie between the 25th to 75th percentiles or outside them, or between the 40th to 60th percentiles and outside the 10th and 90th percentiles (sometimes, these questions must be phrased in ways to avoid suggesting the answer)

- establishing a strong facilitator who ensures each participant must individually put his evidence on the table and justify it (the facilitator must use his/her judgment on when to push the participants, rather than going through a long and tedious checklist)

- being careful when assessing parameters that are not directly observable (the distribution is supposed to reflect the expert's evidence concerning a particular parameter; if the expert has little direct experience with the parameter, it can be difficult to justify an informative prior distribution)

Typically, the facilitator is an HRA/PRA analyst, who has gained expertise in elicitation at least to the level of understanding described in this section and is accustomed to interpreting qualitative evidence as probabilities.

Controlling for Unintentional Bias

One of the most important concerns associated with the use of a consensus expert judgment process is that of unintentional bias. In the subjective process of developing probability distributions, strong controls are needed to prevent bias from distorting he results (i.e., to prevent results that don't reflect the team's state of knowledge). Perhaps the best approach is to thoroughly understand how unintended bias can occur. With that knowledge, the facilitator and the experts can guard against its influence in their deliberations. A number of issues need to be considered.

A number of studies [e.g., Refs. 20–22] present substan ial evidence that people [both naive analysts and subject matter (domain) experts] are not naturally good at estimating probability (including uncertainty in the form of probability distributions or variance). For example, Hogar h [Ref. 21] notes that psychologists conclude hat man has only limited information processing capacity. This in turn implies that his perception of information is selective, that he must apply heuris ics and cognitive simplification mechanisms, and that he processes information in a sequential fashion. These characteristics, in turn, often lead to a number of problems in assessing subjec ive probability. Evaluators often:

- ignore uncertainty (this is a simplification mechanism); uncertainty is uncomfortable and complicating, and beyond most people's training

- lack an understanding of the impact of sample size on uncertainty. Domain experts often give more credit to their experience than it deserves (e.g., if they have not seen it happen in 20 years, they may assume it cannot happen or that it is much more unlikely han once in 20 years)

- lack an understanding or fail to think hard enough about independence and dependence

- have a need to structure the situation, which leads people to imagine patterns, even when here are none

- are fairly accurate at judging central tendency, especially the mode, but tend to significantly underestimate the range of uncertainty (e g., in half the cases, people's es imates of the 98% intervals fail to include he true values) and are influenced by beliefs of colleagues and by preconceptions and emotions

- rely on a number of heuristics to simplify the process of assessing probability distributions; some of these introduce bias into the assessment process

Examples of this last area include:

- Representativeness: People assess probabilities by the degree to which they view a known proposi ion as representative of a new one. Thus stereotypes and snap judgments can influence their assessment. In addition, representativeness also ignores the prior probability (i.e., what their initial judgment of the probability of the new proposition would be, before considering the new evidence—in his case their assumption of the representativeness of the known proposition). Clearly the prior should have an impact on he posterior (revised) probability, but basing our judgment on similarity alone ignores that point. This also implies that representativeness is insensitive to sample size (because they jump to a final conclusion, based on an assumption of similarity alone).

- Availability: People assess the probability of an event by the ease with which instances can be recalled. This availability of the information is confused with its occurrence rate. Several associated biases have been observed:

    - biases from the retrievability of instances-recency, familiarity, and salience

    - biases from the effectiveness of a search set-the mode of search may affect the ability to recall

    - biases of imaginability-the ease of constructing inferences is not always connected with the probability

- Anchoring and Adjustment: People start with an initial value and adjust it to account for other factors affec ing the analysis. The problem is that it appears to be difficult to make appropriate adjustments. It is easy to imagine being locked to one's initial estimate, but anchoring is much more sinister than that alone. A number of experiments have shown that even when the initial estimates are totally arbitrary, and represented as such to the par icipants, the effect is strong. Two groups are each told that a starting point is picked randomly just to have a place to work from. The one given the higher arbitrary starting point generates a higher probability. One technique found to be helpful is to develop estimates for the upper and lower bounds before addressing most likely values.

Lest we agree prematurely that people are irretrievably poor at generating subjec ive estimates of probability, it is signficant to realize that many applications have been successful. Hogarth [Ref. 21] points out that studies of experienced meteorologists have shown excellent agreement with actual facts. Thus, an understanding is needed of what techniques can help make good assessments. In addi ion, in his comments published with the Hogarth paper, Edwards observes that humans use tools in all tasks, and tools can help us do a very good job in the elicitation process.

Winkler and Murphy [Ref. 23] make a useful distinction between two kinds of expertise or "goodness. "Substantive" expertise refers to knowledge of the subject matter of concern. "Normative" expertise is the ability to express opinions in probabilistic form. Hogarth [Ref. 21] points out that the subjects in most of the studies were neither substantive nor normative experts. A number of studies have shown that normative experts (whose domain knowledge is critical) can generate appropriate probability distributions, but that substantive experts require significant training and experience, or assistance (such as provided with a facilitator), to do well.

For the estimates of uncertainty to be meaningful, they must be based on a formalism that accounts for collected experimental and experiential evidence. When such information is lacking or incomplete, it must be systematically developed from all available evidence, including the judgments of the experts. The formalisms for describing uncertainty and its relevancy when performing expert elicitation are discussed in Appendix B.

3.8.2.3 Eliciting Expert "Evidence"

The expert elicitation approach has both its critics and supporters. Appendix B provides a summary of the relevant issues regarding the viability of using expert elicitation, and the reader is encouraged to review that information.

In order to avoid the most significant pitfalls of using expert elicitation, as discussed in Appendix B, it is important to observe that which makes experts "expert" is not their opinions but their knowledge, experience, experiments, etc.— in short, their evidence. Therefore, instead of asking the experts for their opinions, what is needed is an elicitation of their evidence.

This is an important subtlety and one that is useful in performing an appropriate elicitation to estimate the HEPs that need to be assessed. What this does is focus the elicitation of the experts to address the following question:

- "What evidence and information do you have that is relevant to estimating the HEP for the given context?"

rather than the following:

- "What is your best estimate for the HEP?"

- "What is your state of confidence about the HEP estimate?"

Then, the HRA/PRA analyst (likely acting as the facilitator) takes the lead role in converting the collective evidence (data) into a probability [i.e., the HEP with an uncertainty distribution, albeit still with the input of the experts (the highly recommended practice) although this could be done just by the HRA analyst using the elicited evidence]. The conversion into a probability is something the HRA/PRA analyst is much more comfortable and experienced in doing than are most if not all the experts and so his/her guidance in this part of the process is important.

3.8.3 *Guidance*

Step 8 and its formal, controlled elicitation, is carried out by performing 10 tasks:

1. Gather the experts.

2. Thoroughly explain the context and the HFE/UA.

3. Elicit relevant evidence from the experts.

4. Guide the subsequent discussion.

5. Confirm the evidence.

6.   Elicit each expert's HEP.

7.   Construct a consensus HEP.

8.   Repeat previous tasks for each HEP to be assessed.

9.   Perform a sanity check of the estimated HEPs.

10.   Document the quantification.

Note about iteration back to previous steps:

While Step 8 is not intended to be a complete check of everything that has been done in previous steps in the ATHEANA process, it may become apparent that a previous step in the ATHEANA process needs to be revisited because of the experts' input. For instance, because of the discussions during the elicitation it may become desirable to redefine an existing HFE/UA, or to define a new HFE/UA, or to consider a new context, or to reassess a previous judgment about the likelihood of recovery, or to revisit the previous screening of a deviation scenario. If necessary, iteration on the appropriate step(s) should be performed either before doing the expert elicitation or even during the elicitation if that is possible. The important point is to document any necessary changes so that it is clear what the inputs actually are for the elicitation.

3.8.3.1   Task 8.1: First, Gather the Experts, If Possible, in One Room[10]

This is important. The kind of interaction needed to make the process work as described does not seem to work long distance—via telcon, email, or letter. It has as much to do with body language and facial expression as with the actual words communicated. There is a feeding on each other's ideas and challenges that is essential in pulling out a complete sharing of information.

3.8.3.2   Task 8.2: Explain Thoroughly the Context and HFE/UA Being Addressed

Led by the facilitator/HRA/PRA analyst(s), make sure that everyone understands exactly what the definition of the HFE/UA is and the context for which the HEP for that HFE/UA is to be estimated. In other words, we have to make sure that everyone understands the philosophy and purpose of the analysis being done and the precise role and meaning of the HFE/UA in this model. Much interchange should be encouraged at this point. This step must be done well or there will be confusion later, and the participation of the experts will be halfhearted.

---

[10]   Kaplan, Bley, and Johnson [Ref. 24] have found that groups of 5–10 can be handled nicely. The USNRC SSHAC report [Ref. 25] lays out different levels of formalism for different kinds of problems. There is no reason why even a single analyst, who must address uncertainty in an analysis should not develop the required normative skills and be formal in documenting the available evidence and using it to construct a probability distribution for the parameter of interest. He/she may need help in gathering evidence, but still do the evaluation himself (herself).

Reflecting this general elicitation guidance on the HRA, this task involves an open discussion, summarizing all that has been learned from the qualitative portion of the ATHEANA process, in order to prepare for estimating the HEP associated with the HFE/UA being addressed. It is a way to "get all the experts up to speed and on the same page" about what needs to be considered in making the HEP estimate. Ideally, the facilitator or HRA or PRA analyst(s) will describe the scenario in terms of the expected plant conditions and the HFE/UA being examined, and summarize information on any previously identified likely influencing PSFs. The goal here is to be as factual as possible about what has been learned in the qualitative analysis leading up to the elicitation and yet avoid biasing the expert panel such as by emphasizing one aspect more than another.

In addition, if some of the experts participating in the elicitation were not involved in the screening of any deviation scenarios related to the HFE/UA being addressed (e.g., in Steps 6 or 7), then it will also be useful to discuss those related scenarios that were screened to ensure that everyone agrees that they were appropriately screened. These discussions may also help clarify the context for the HFE/UA and scenario being quantified.

Note about dependencies among human actions in a scenario:

One point is worth highlighting at this stage. If, for the relevant PRA scenario and the context being considered, it is known that one or more other human actions are assumed to be successful or failed as part of the entire chain of events for the scenario, this should be covered in describing the context. Clearly, what the operating crew has done earlier in a scenario could have an important bearing on what they do later. It is part of the context and one of the factors that can influence performance. Although Section 3.5.2.3 provides a discussion of factors to consider in assessing across scenario dependencies, ATHEANA does not have a detailed dependency model with a specific set of rules to follow. Rather, it relies on the experts to evaluate what a failure or success earlier in a scenario might imply about the crew's understanding of what was happening, how that might affect performance later, and how some effects might be tempered by aspects such as additional cues and system feedback (see Section 3.5.2.3). Thus, possible dependencies among the human events should be addressed during the discussion of context to be sure the experts understand the nature of these other human successes or failures and their role as part of the context. In that way, it can be ensured that the intent of the elicitation is to estimate the HEP for the HFE/UA being assessed, given one or more other human actions have succeeded or failed as part of the scenario.

3.8.3.3   Task 8.3: Elicit the Evidence

After the meaning of both the HFE/UA and the relevant context (the nominal context if that is the one being addressed or one of the EFCs if that is being addressed) is clear, the facilitator then puts to the group the question: "What evidence do we have that is relevant to the likelihood of the crew (operator) making the error described by the HFE/UA?"

3.8.3.4   Task 8.4: Guide the Discussion

In the ensuing discussion, the facilitator guides the group to clarity and agreement on the meaning of each item of evidence. The facilitator writes these down in his/her notes and on a visual medium such as a blackboard or projected screen able to be seen by all the members of the group.

Here the discussion continues from, and is an expansion of, the introduction in Task 8.2. Each member of the group is called upon to put all their evidence on the table. Table 3.8-1 provides examples of the information that comes from the qualitative portion of the ATHEANA process (from previous steps) and would be summarized, reviewed, discussed, and added to openly among the experts, but led by the facilitator.

Table 3.8-1. Examples of Information Expected to be Discussed

Information Type	Examples
Plant conditions & behavior for scenario/context	Thermal-hydraulic conditions as a function of time, expected plant indications as a function of time, system/equipment operations, expected operator actions.
Critical plant functions for accident mitigation	Specific equipment operation, requirements for operator action, possible operator recovery actions.
Operating crew characteristics (i.e., crew characterization)	Crew structure, communication style, emphasis on crew discussion of "big picture", behaviors observed in simulator exercises and/or identified by training staff.
Features of procedures	Structure, how implemented by operating crews, opportunities for "big picture" assessment and monitoring of critical safety functions, emphasis on relevant issue (e.g., ensure injection), priorities, any potential mismatches especially with deviation scenarios.
Relevant informal rules	Experience, training, practice, ways of doing things - especially those that may conflict with informal rules or otherwise lead operators to take inappropriate actions.
Timing	Plant behavior and requirements for operator intervention versus expected timing of operator response in performing procedure steps, etc.; input from training staff and results of simulator exercises; based upon perceived needs of the PRA, multiple times or time frames may need to be considered for each HFE/UA.
Relevant vulnerabilities	Any potential mismatches between the scenarios and expected operator response with respect to timing, formal and informal rules, biases from operator experience and training, and so forth.
Performance-shaping factors	Those deemed associated with or triggered by the relevant plant conditions and including whether they are positive or negative influences and the strengths of their influence on operator performance for the context (e.g., missing or misleading indications, complex situations, timing mismatches and delays, procedural ambiguities, workload, and human-machine interface concerns).
Recovery potential	Possible recovery actions if the initial error should be made. Confirm/change that found during Step 7 such as cues for doing so, time available before undesired consequences, staff resources for doing so, and so forth.

The goal of the ensuing open discussion is not necessarily to achieve consensus opinions regarding the results of these discussions, but rather to advance the understanding of all the experts through the sharing of distributed knowledge and expertise. In each case, the scenario and its context and the HFE/UA in question are described and the vulnerabilities and strong points associated with taking the right action for the context are discussed openly among the team.

Specifically, the facilitator must perform the following tasks:

(1) Collect from the team (or make assumptions about) any additional information that is not already available and that is needed to describe and define the HFE/UA and associated context.[11]

(2) Ensure that all members of the team review all information for clarity, completeness, accuracy [some of the team may not agree with assumed facts (e.g., plant behavior)].

(3) Interpret and prioritize all information with respect to relevance, credibility, and significance.

Item 3 is especially important if any of the following conditions exist:

• There are conflicts between information sources.

• Information is ambiguous, confusing, or incomplete.

• Information must be extrapolated, interpolated, etc.

3.8.3.5 Task 8.5: Confirm the Evidence

In this step, the facilitator confirms that the evidence provided is sufficiently complete, relevant, and clear in order to proceed with the quantification phase of the elicitation. When this step is done, there is on the blackboard or projected screen (and captured in some form of documentation) a complete list of understood and agreed-upon evidence items: E1, E2... Em. The facilitator needs to make sure here that the experts and the facilitator are satisfied that this list captures the total experience and information of the group.

3.8.3.6 Task 8.6: Elicit Each Expert's HEP and its Distribution

Given all the above knowledge (evidence), it is in this step that the qualitative information is first converted into a quantitative estimate of the error rate (i.e., the HEP), including any potential to recover from the initial error should that error be made. That is, the HEP is to account for recovery of any initial error to the extent appropriate. There are a number of activities that need to be performed in implementing Task 8.6.

---

[11] This process of collecting the evidence, requires that the facilitator be alert to the issues raised in the previous text boxes on "The Facilitator" and "Controls for Unintentional Bias." Specifically, the facilitator must insist that every member of the group put his or her own evidence and experience on the table, not just say, "I agree with what Charlie said." The facilitator must read the body language of the participants and pursue any indications that information or disagreement is being withheld. For example, "Sally, you don't look comfortable with this characterization of the plant conditions, what are your concerns?" or "John, you seem concerned that we are making assumptions that might not be consistent with the information available to the operator. As an operator, what screens would you actually be focused on? Are there Shift Supervisors in other crews that might be selecting different information?"

In this step, the facilitator, with the assistance of the experts, puts forth two questions that progressively move the entire group from a qualitative evaluation to a quantitative estimate of the HEP and its uncertainty distribution:

(1)   Given all the relevant evidence, how difficult or challenging is the action of interest for the scenario/context and why?

(2)   Hence, what is the probability distribution for the HEP that best reflects this level of difficulty or challenge considering uncertainty?

The important point to recognize here is that particularly question #2 above, although put to the group, really falls within the domain of the facilitator/HRA/PRA analyst, not the subject matter experts. These experts have already contributed their knowledge in the form of the collective evidence. The representation or mapping of this evidence into an HEP is an operation best managed by the facilitator/analyst. The facilitator/analyst could perform this transformation of the evidence into an HEP by himself/herself such as if people resources and time are severely limited (i.e., skip most or all the remaining parts of the elicitation carried out with the group). However, if possible, arriving at the estimate for the HEP is best done with the assistance and the agreement of the experts.

Pose the First Question

By this time in the elicitation process, the experts are likely to have come to individual impressions about their answers to question #1. The facilitator should begin this elicitation step by posing the first question to the group and having the experts individually provide their response and rationale. While the experts should be allowed to offer their views in whatever way is best for each of them, the general form of the answer being sought is:

> "The action will be (easy, hard, extremely difficult, etc.) for the crew (or individual operator if that is appropriate) because..."

Said another way, this is each expert's overall qualitative conclusion about the difficulty or level of challenge for the action in light of all the evidence and given the plant conditions germane to the scenario and its context, including the most influencing PSFs as well as considering the likelihood of recovering from an initial error.Following each individual response, an open discussion is then held (led by the facilitator), focused on the differences among the responses. The purpose of this discussion is two-fold. First, it is intended to find areas where experts are not using some of the evidence, or misinterpreting the evidence, or using other evidence not previously discussed, so that these differences can be shared or otherwise rectified to the extent possible. Second, if possible, a consensus overall opinion should be reached as to the level of difficulty or challenge for the action of interest and the major reasons (i.e., most influencing factors) for that level of difficulty or challenge. While a consensus is sought, it is recognized that "reasonable people may choose to disagree" and if that is the case, the disagreements should be noted and later factored into the HEP estimate.

Pose the Second Question and Discuss Useful Probability Calibration Points

At this point, and with knowledge of the overall conclusion reached above, the facilitator poses the second question about the probability distribution for the HEP. In starting out to develop the distribution with the experts help, limited applications of ATHEANA have found it useful to first provide a calibration mechanism for the experts to begin to interpret their qualitative conclusions into a probability.

This is first done by having the experts try to imagine how many times they would expect crews (or an individual operator if that is more appropriate for the action of interest) to commit the HFE/UA (such as in a simulation of the scenario and its context) as a reflection of the level of difficulty or challenge that has been previously expressed. The following table often proves helpful in these initial evaluations, until the experts begin to develop a sense of the meaning of the probability values. While it is sometimes recommended that experts be limited to a few specific choices, we have found that they quickly begin to demand more flexibility in their assignments, which is encouraged. Table 3.8-2 provides a suggestion for this initial calibration.

Table 3.8-2. Suggested Set of Initial Calibration Points for the Experts

Circumstance	Probability	Meaning
The operator(s) is "Certain" to fail	1.0	Failure is ensured. All crews/operators would not perform the desired action correctly and on time.
The operator(s) is "Likely" to fail	~ 0.5	5 out of 10 would fail. The level of difficulty is sufficiently high that we should see many failures if all the crews/operators were to experience this scenario.
The operator(s) would "Infrequently" fail	~ 0.1	1 out of 10 would fail. The level of difficulty is moderately high, such that we should see an occasional failure if all of the crews/operators were to experience this scenario.
The operator(s) is "Unlikely" to fail	~ 0.01	1 out of 100 would fail. The level of difficulty is quite low and we should not see any failures if all the crews/operators were to experience this scenario.
The operator(s) is "Extremely Unlikely" to fail	~ 0.001	1 out of 1000 would fail. This desired action is so easy that it is almost inconceivable that any crew/operator would fail to perform the desired action correctly and on time.

Besides the above initial calibration points, the group should share their experiences that may be useful in assigning an HEP that is reflective of the level of difficulty for the human action and the reasons therefore. These can come from actual training and simulator experiences, real plant events, other industry events, and even outside and every day experiences that provide insights into how often things go wrong vs. the number of opportunities for different levels of overall difficulty associated with different actions.

Armed with these aids to help the experts turn qualitative judgments into probability estimates, the facilitator asks the experts to make their own judgment for the estimate for the HEP in question. However, in doing so, a distribution representing uncertainty in the HEP estimate rather than a single value is to be provided by the experts. This leads to the next activity to provide the answer that is desired for question #2.

Developing a Distribution for the HEP from Each Expert

In coming up with a distribution for the HEP, the experts are to be reminded that they cannot be exactly sure how all the characteristics of the scenario and its context will influence performance. That is, there is epistemic uncertainty as a result of this lack of knowledge. The facilitator/analyst may need to explain to the experts the use of probability curves as expressions of state of confidence and the use of the probability density function (pdf) as a means to portray such uncertainty in an analysis. Its use can be demonstrated, for instance, by having the facilitator/analyst work with the group to offer their distribution representing how many people in the group have certain years of experience on the job, or fit within certain age groups, based on known evidence as well as such evidence as looking at the people to guess their ages or years of experience. Characteristics of the resulting distribution such as the mode, median, mean, and extremes, and what they mean, can be discussed. With these concepts in mind, and to avoid certain unintentional biases by addressing the extremes of the distribution first, the experts are instructed to individually perform the following tasks:

- First assess what will be interpreted as the 99th percentile value for the HEP distribution. This can be explained to be that value for the HEP that each expert believes the error rate cannot be any higher considering all the evidence and armed with the probability calibration points discussed earlier. In other words, given the evidence, what is the worst or the highest the probability of failure could be for UA.

- Similarly, the experts are asked to individually assess that value for the HEP for which the error rate cannot be any lower. That is, what is best or the lowest the failure probability could be. This will be interpreted as the 1st percentile value.

- With the 1st and 99th percentiles identified, the experts are then asked to individually identify that value for the HEP that each expert believes to be the most likely appropriate value for the HEP (the mode).

- Finally, the experts are asked to provide a "rough shape" for the pdf covering these values that is reflective of their degree of uncertainty in the HEP. For instance, one expert may believe that while the two extremes he/she provided are possible, that expert judges that the shape of the distribution is highly peaked about the mode and drops off quickly on both sides indicating a high degree of belief that the mode value is the most appropriate value for the HEP. Another expert may have considerable uncertainty in the most appropriate value as might be illustrated by a broad and relatively flat distribution shape. Still others may believe that the shape is skewed to one side or the other of the mode according to their belief as to which side of the mode is most representative of the HEP.

At this point, each expert is asked by the facilitator to independently provide his/her distribution including the three estimates and the general shape for the HEP being evaluated. Once all the expert's values and approximate shapes are recorded and shown to the group, each expert is asked to describe the reasons why he or she chose the values and shape presented. An open discussion should be led by the facilitator allowing the experts to express their views and possibly affect other experts to want to change their estimates in light of this shared discussion. During this process, the facilitator is reminded of the roles of the facilitator and the need to control unintentional biases as covered in section 3.8.2.1 above. The facilitator should work with each expert to test the values and shape of the presented distributions. For instance, the shape/skewness of the distribution can be examined by having the facilitator note where, approximately, the 25th percentile and 75th percentile values appear to be and asking the expert whether the implications of those values and the relative areas under the curve to each side of these values are really what the expert intended. Other points on the distribution can be similarly examined. Questions can be used such as "are you saying that you would wager that there is a 25% chance the HEP is above (or below) value X?" or "are you saying that value Y is representative of that value for the HEP whereby there is a 50:50 chance that the HEP lies on either side of value Y?" What the facilitator/analyst is doing is helping each expert better define their distributions based on their qualitative judgements and testing the values that represent, for instance, the 25th percentile, the 75th percentile, the median, etc., based on the distribution shape.

During this discussion, the facilitator may also want to guide the group back to Tasks 8.4 or 8.5 or the earlier part of Task 8.6 as may be necessary to clear up any inappropriate interpretations leading to the different estimates. If any strong aleatory sources of uncertainty are being manifested in this discussion, they may need to be pulled out as separate influences thus contributing to a new context that needs to be developed and have its corresponding HEP assessed. Based on this discussion, any revisions in the individual estimates should then be allowed to be made, and recorded by the facilitator. It should be noted that the experts are not required to modify their original estimates.

3.8.3.7   Task 8.7: Construct a Consensus HEP and its Distribution

With this information, the facilitator should lead the group into building a consensus distribution that is reflective of the different values and shapes offered by the experts. There is no demonstrably correct or best way to combine the estimates — with algebraic averaging high outliers can dominate results; with geometric averaging low outliers can dominate. Given the "rough shaping" of the distributions anyway, it will most likely be sufficient to develop a consensus distribution on the basis of approximating a distribution that encompasses the extremes of the values provided and accounts for the shapes as well as the closeness and number of occurrences of the same, or similar values. The resulting approximation of a consensus distribution may be quite broad especially if the experts disagree as to the level of difficulty associated with the human action as reflected in their estimates. In such a case, the consensus distribution is demonstrating that the group, as a group, has a large uncertainty about the HEP. Otherwise, the consensus distribution could be quite tight demonstrating a small uncertainty if the individual values are all nearly the same.

In the end, the group needs to agree that the consensus distribution is appropriate and reasonably representative of the uncertainty in the HEP based on all the experts' judgments. Similar testing by the facilitator is recommended of individual values implied by the consensus shape as was done with each expert's individual input earlier.

3.8.3.8 Task 8.8: Repeat Elicitation Tasks 8.1 Through 8.7 for All the HEPs To Be Estimated

The same process is followed for each of the HEPs to be estimated until all the HFEs/UAs are addressed across all the contexts to be considered. In general, it will probably be most efficient to quantify each UA and EFC(s) associated with a given HFE before proceeding to a new HFE. However, the issue of combining the probabilities of multiple UAs and EFCs for a given HFE is not addressed until Step 9.

3.8.3.9 Task 8.9: Perform a Sanity Check of the Estimated HEPs

Once all the HEPs have been estimated, the group should perform a sanity check of the estimated HEPs for consistency of the quantifications. This may be done immediately or once a summary document has been prepared by the facilitator or his/her designee (see Task 8.10 below) that captures all that was done during the elicitation including the consensus probability curves, the evidence items having the most influence on these curves, and the reasoning connecting them.

The estimates should be reviewed relative to each other to check their reasonableness considering the actions and the scenarios and contexts. It should be evident that those HFEs/UAs with the highest HEPs make sense considering the plant conditions and PSFs involved while those with the lowest also make sense given the relative ease of these actions for the circumstances involved. In other words, the probabilistic ordering of the HEPs should be reasonable given the different actions and contexts involved.

If necessary, evaluations may have to be revisited/redone if the sanity check suggests inconsistencies among the HEPs.

3.8.3.10 Task 8.10: Document the HEP Elicitation and Sanity Check

Throughout the elicitation process, including the sanity check, detailed notes and other records of each HEP elicitation should be made so that a writeup on each HEP estimate can be produced. These writeups should be able to provide the experts as well as others not involved in the elicitation a detailed summary of the results of each task for each HEP so that the origin of the consensus distribution for each HEP can be clearly traced and understood. Later, when the estimates are incorporated into the PRA or similar risk framework (see Step 9), the effects of these estimates on the overall risks being evaluated should be further described. Any comments from the experts or others qualified to comment on these writeups should be addressed to finalize the documentation of the estimated HEPs.

3.8.4 *Output*

The major outputs of this step include the following:

(1) The HEP estimate, including its uncertainty distribution, for each HFE/UA analyzed for each relevant scenario and context. These HEP estimates are incorporated into the PRA (or other similar risk framework) as appropriate (see Step 9).

(2) Detailed writeups and supportive notes and other records documenting each HEP estimate and the basis for the estimate. A good description of the factors driving performance should be provided.

Additionally, this process has many side benefits besides the production of HEPs used to base decisions. For one, the experts are forced to become explicit about their evidence. Each expert learns what the experience and information of the others are. In the course of becoming explicit, each item of evidence is thoroughly discussed, examined, challenged, and compared for consistency with the other items of evidence; interpretations are debated; semantics are clarified; and fine distinctions are drawn. The resulting agreed-upon information base relevant to each HFE/UA, in itself, is a very useful result.

Furthermore, because this is a probabilistic analysis, the HFEs/UAs and their corresponding HEPs will likely be related directly or indirectly to the occurrence of PRA scenarios. Hence, during the discussions, while the group is focusing on a specific scenario, ideas for ways in which to eliminate this scenario or to make it less likely, will likely surface. Someone will say, "If that's an area of concern, we can do such and such or change this and that." This idea will spark further ideas, leading to definite proposals for changing design and procedures to reduce risk.

3.9   Step 9: Incorporate Each HFE/UA and Corresponding HEP Into the PRA

3.9.1   *Purpose*

The purpose of this step is to incorporate the results of the ATHEANA analysis into the PRA. Generally, the process is not any different than would normally be done using any HRA method. However, because ATHEANA potentially covers a range of contexts and UAs, and may model EOCs, a few considerations need to be addressed.

3.9.2   *Guidance*

As discussed in section 2.3.3, there are two basic ways to incorporate the results of the ATHEANA analysis into the PRA. The first way involves maintaining the PRA logic model and original defined HFE intact. The frequency of the accident sequence excluding the HFE is determined the typical way based on the probabilities of the successes and failures associated with the sequence including the frequency of the initiating event. The HEP for the HFE would be the HEP distribution (covering epistemic uncertainty) obtained from applying the quantification process to each combination of EFC (including the nominal context) and UA modeled for a given HFE, and summing the results per the ATHEANA quantification formula. In other words, the HEP distributions for each UA and EFC combination associated with a given HFE, would be convolved and the resulting HEP distribution would be used for the HFE in the PRA model. If a single UA, with a single context (e.g., the nominal scenario) is all that is modeled for a given HFE, then the HEP distribution for that combination is all that will be used for that HFE in the model. The second way is to expand the original PRA modeled sequence to explicitly reflect the different contexts and the specific UAs including any EOCs for each context. This might be done, for instance, in either the event trees by adding top events or fault trees by adding basic events. The HEPs (epistemic distributions) would then be applied to the UAs for each of their contexts represented in the PRA model as appropriate. Section 2.3.3 provides a discussion and Figure 2.3-1 provides an illustration of both ways to incorporate the results into the PRA.

In most cases, whether the analysts use a single HFE in the logic model to represent the range of contexts and UAs or explicitly represent the different combinations by adding new events to the model is simply a matter of preference. One obvious tradeoff is that by adding new events to the trees, the conditions being modeled are explicitly incorporated into the model structure, but the model will become much larger. By explicitly representing the different combinations by adding new events, the PRA software used to solve the model performs the appropriate calculations involved in combining the different HEP results.

However, there are at least a couple of cases where it may be more appropriate to create new events and alter the logic model. The first is when one of the UAs is an EOC. Usually, it will take very different contexts to lead to an EOC than to an EOO, and therefore the dependencies on events downstream in the PRA logic model could be very different. Similarly, if there is something about the different contexts that could lead to a given EOO (i.e., that lead to the same UA) that could affect what the operators do downstream in different ways, then it may also facilitate the dependency analysis downstream to have the different conditions represented explicitly in the event or fault trees. Such considerations are relevant to later events in trees and for the cutset or sequence recovery analysis. Furthermore, if the original modeled HFE is not altered when adding an EOC, then the HEP for the HFE is made up of an EOC contribution and a EOO contribution and there is no explicit distinction apparent in the PRA. In such cases, it is highly recommended that EOCs be modeled separately from EOOs with explicit events and logic as appropriate, so that the EOO and EOC forms of human failure are clearly distinguishable.

As a final note, remember that if multiple human failures in the same sequence are not foreseen during the initial quantification of the various UAs and their contexts, then as with any PRA/HRA methodology, there will be an obligation of the analysts to identify such combinations once the PRA is initially "solved" and the human error combinations can be readily identified. Based on this information, HEP evaluations may have to be revisited/redone if the results of these evaluations are potentially significant contributors to the risk and sufficiently strong dependencies are considered to likely exist among certain HFE/UAs. And as noted in Section 8, reasonableness checks of the revised HEPs should be performed again.

3.9.3   *Output*

The output of this step is the incorporation of the results of the ATHEANA analysis into the PRA in the preferred manner and an evaluation of the results for any potential dependencies that may have been missed.

4. REFERENCES

[1] NUREG-1624, "Technical Basis and Implementation Guidelines for A Technique for Human Event Analysis (ATHEANA)," Rev. 1, U.S. Nuclear Regulatory Commission, Washington, DC, May 2000.

[2] ASME RA-Sb-2005, "Standard for Probabilistic Risk Assessment for Nuclear Power Plant Applications," Addendum B to ASME RA-S-2002, American Society of Mechanical Engineers, New York, NY, December 30, 2005.

[3] NUREG-1792, "Good Practices for Implementing Human Reliability Analysis (HRA)," U.S. Nuclear Regulatory Commission, Washington, DC, April 2005.

[4] Julius, J.A., E.J. Jorgenson, A.M. Mosleh, and G.W. Parry, "A Procedure for the Analysis of Errors of Commission During Non-Power Modes of Nuclear Power Plant Operation," *Reliability Engineering & System Safety*, 53:139–154, Elsevier, Amsterdam, The Netherlands, 1996.

[5] Julius, J.A., E.J. Jorgenson, G.W. Parry, and A.M. Mosleh, "A Procedure for the Analysis of Errors of Commission in a Probabilistic Safety Assessment of a Nuclear Power Plant at Full Power," *Reliability Engineering & System Safety*, 50:189–201, Elsevier, Amsterdam, The Netherlands, 1995.

[6] Macwan, A., and A.M. Mosleh, "A Methodology for Modeling Operator Errors of Commission in Probabilistic Risk Assessment," *Reliability Engineering & System Safety*, 45:139–157, Elsevier, Amsterdam, The Netherlands, 1994.

[7] Reer, B., V.N. Dang, and S. Hirschberg, "The CESA Method and its Application in a Plant-Specific Pilot Study on Errors of Commission," *Reliability Engineering & System Safety*, 83:187–205, Elsevier, Amsterdam, The Netherlands, 2004.

[8] Wakefield, D., G.W. Parry, G. Hannaman, and A. Spurgin, "SHARP1: A Revised Systematic Human Action Reliability Procedure," EPRI TR-101711, Tier 2, Electric Power Research Institute, Palo Alto, CA, December 1992.

[9] Reason, J., *Human Error*, Cambridge University Press, New York, NY, 1990.

[10] Wakefield, D.J., "Application of the Human Cognitive Reliability Model and Confusion Matrix Approach in a Probabilistic Risk Assessment," *Reliability Engineering & System Safety*, 22:295–312, Elsevier, Amsterdam, The Netherlands, 1988.

[11] Potash, L.M., M. Stewart, P.E. Dietz, C.M. Lewis, and E.M. Dougherty, Jr., "Experience in Integrating the Operator Contributions in the PRA in Actual Operating Plants," in *Proceedings of the ANS/ENS Topical Meeting on Probability Risk Assessment, Port Chester, NY, September 1981*, pp. 1054–1063.

[12] Swain, A.D., and H.E. Guttmann, NUREG/CR-1278, "Handbook of Human Reliability Analysis with Emphasis on Nuclear Power Plant Applications," U.S. Nuclear Regulatory Commission, Washington, DC, 1983.

[13] Kolaczkowski, A., D. Bley, S. Cooper, J. Forester, N. Siu, E. Thornsbury, H. Woods, and J. Wreathall, "Field Test of ATHEANA (A Technique for Human Event Analysis) in Pressurized Thermal Shock Probabilistic Risk Assessments," in *Proceedings of OECD/NEA/CSNI Workshop on "Building the New HRA: Errors of Commission from Research to Application," Rockville, Maryland, May 7–9, 2001*.

[14] Woods, D.D., L.J. Johannesen, R.I. Cook, and N.B. Sarter, "Behind Human Error: Cognitive Systems, Computers, and Hindsight," Crew System Ergonomics Information Analysis Center (CSERIAC), Ohio State University, Wright-Patterson Air Force Base, Columbus, OH, December 1994.

[15] Roth, E.M., R.J. Mumaw, and P.M. Lewis, NUREG/CR-6208, "An Empirical Investigation of Operator Performance in Cognitively Demanding Simulated Emergencies," prepared by Westinghouse Science and Technology Center, for the U.S. Nuclear Regulatory Commission, Washington, DC, July 1994.

[16] Mumaw, R.J., and E.M. Roth, "How To Be More Devious with a Training Simulator: Redefining Scenarios to Emphasize Cognitively Difficult Situations," in *Proceedings of the 1992 Simulation MultiConference: Nuclear Power Plant Simulation and Simulators, Orlando, FL, April 6–9, 1992*.

[17] Perotti, J.W., and D.D. Woods, "A Cognitive Analysis of Anomaly Response in Space Shuttle Mission Control," CSEL 97-TR-02, prepared by Cognitive Systems Engineering Laboratory (CSEL), Ohio State University, for NASA Johnson Space Center, Houston, TX, March 1997.

[18] Woods, D.D., and E.S. Patterson, "How Unexpected Events Produce an Escalation of Cognitive and Coordinative Demands," P.A. Hancock and P.A. Desmond (Eds.), *Stress, Workload, and Fatigue*, Lawrence Erlbaum Associates, Mahwah, NJ, 2000.

[19] Budnitz, R.J., G.E. Apostolakis, D.M. Boore, L.S. Cluff, K.J. Coppersmith, C.A. Cornell, and P.A. Morris, "Use of Technical Expert Panels: Applications to Probabilistic Seismic Hazard Analysis," *Risk Analysis*, 18(4):463–469, Blackwell Synergy, Malden, MA, August 1998.

[20] Tversky, A., and D. Kahneman, "Judgment Under Uncertainty: Heuristics and Biases," *Science*, 185(4157):1124-31, American Association for the Advancement of Science, Washington, DC, September 27, 1974.

[21] Hogarth, R.M., "Cognitive Processes and the Assessment of Subjective Probability Distributions," *Journal of the American Statistical Association*, 70(350):271–294, American Statistical Association, Alexandria, VA, June 1975.

[22] Kahneman, D., P. Slovic, and A. Tversky, *Judgment under Uncertainty: Heuristics and Biases*, Cambridge University Press, Cambridge, England, 1982.

[23] Winkler, R.L., and A.H. Murphy, "'Good' Probability Assessors," *Journal of Applied Meteorology*, 7(5):751–758, American Meteorological Society, Boston, MA, 1968.

[24] Bley, D.C., S. Kaplan, and D.H. Johnson, "The Strengths and Limitations of PSA: Where We Stand," *Reliability Engineering & System Safety*, Elsevier, Amsterdam, The Netherlands, 38(1/2):326, 1992.

[25] Budnitz, R.J., G.E. Apostolakis, D.M. Boore, L.S. Cluff, K.J. Coppersmith, C.A. Cornell, and P.A. Morris, NUREG/CR-6372 (UCRL-ID-122160), "Recommendations for Probabilistic Seismic Hazard Analysis: Guidance on Uncertainty and Use of Experts," prepared by Lawrence Livermore National Laboratory for the U.S. Nuclear Regulatory Commission, U.S. Department of Energy, and Electric Power Research Institute, Washington, DC, April 1997.

APPENDIX A

EXAMPLE FLOWCHART
OF EMERGENCY OPERATING PROCEDURES
FOR A LOSS OF MAIN FEEDWATER SCENARIO

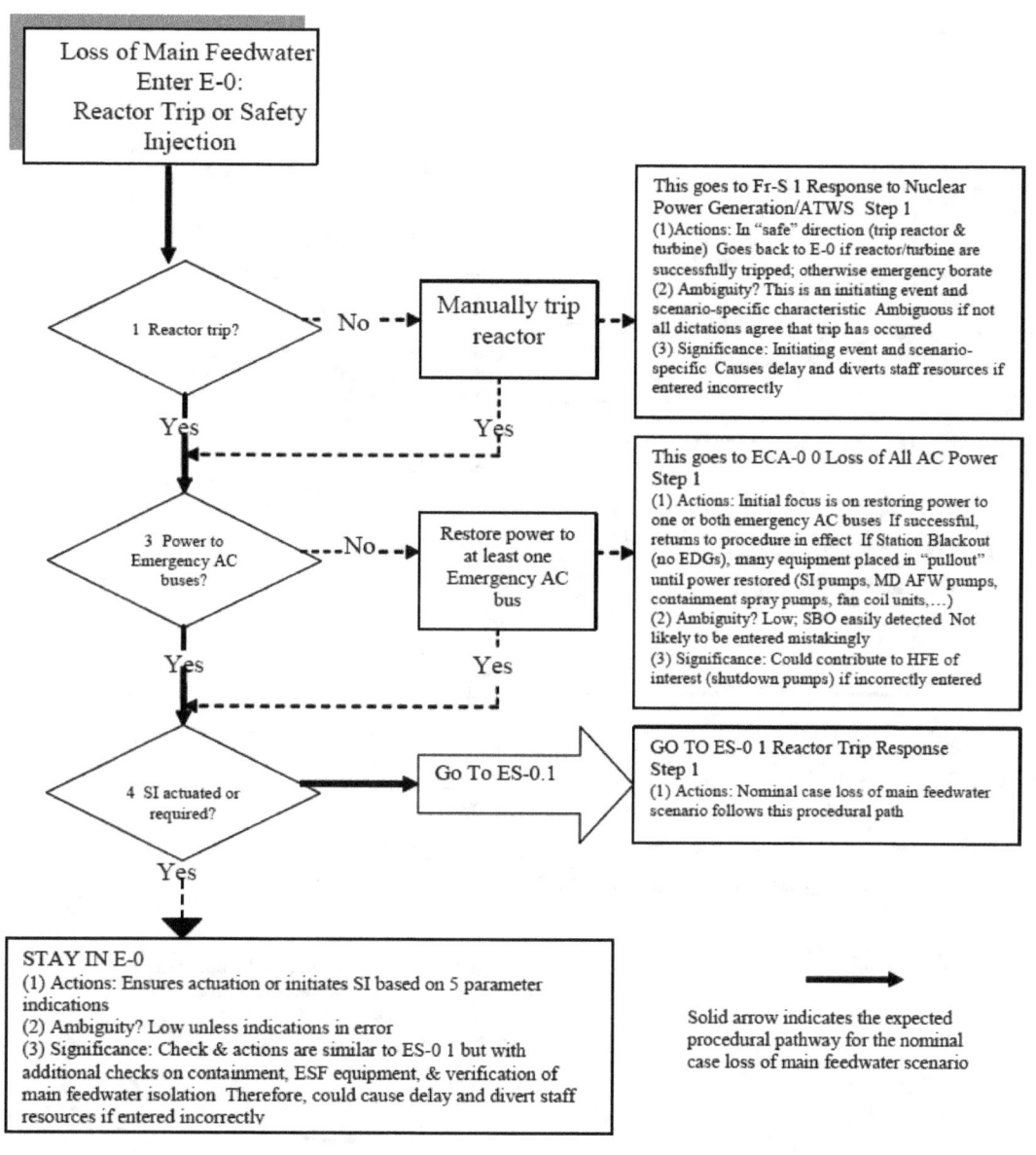

Figure A-1a. EOP Highlights Related to Loss of Main Feedwater Scenario

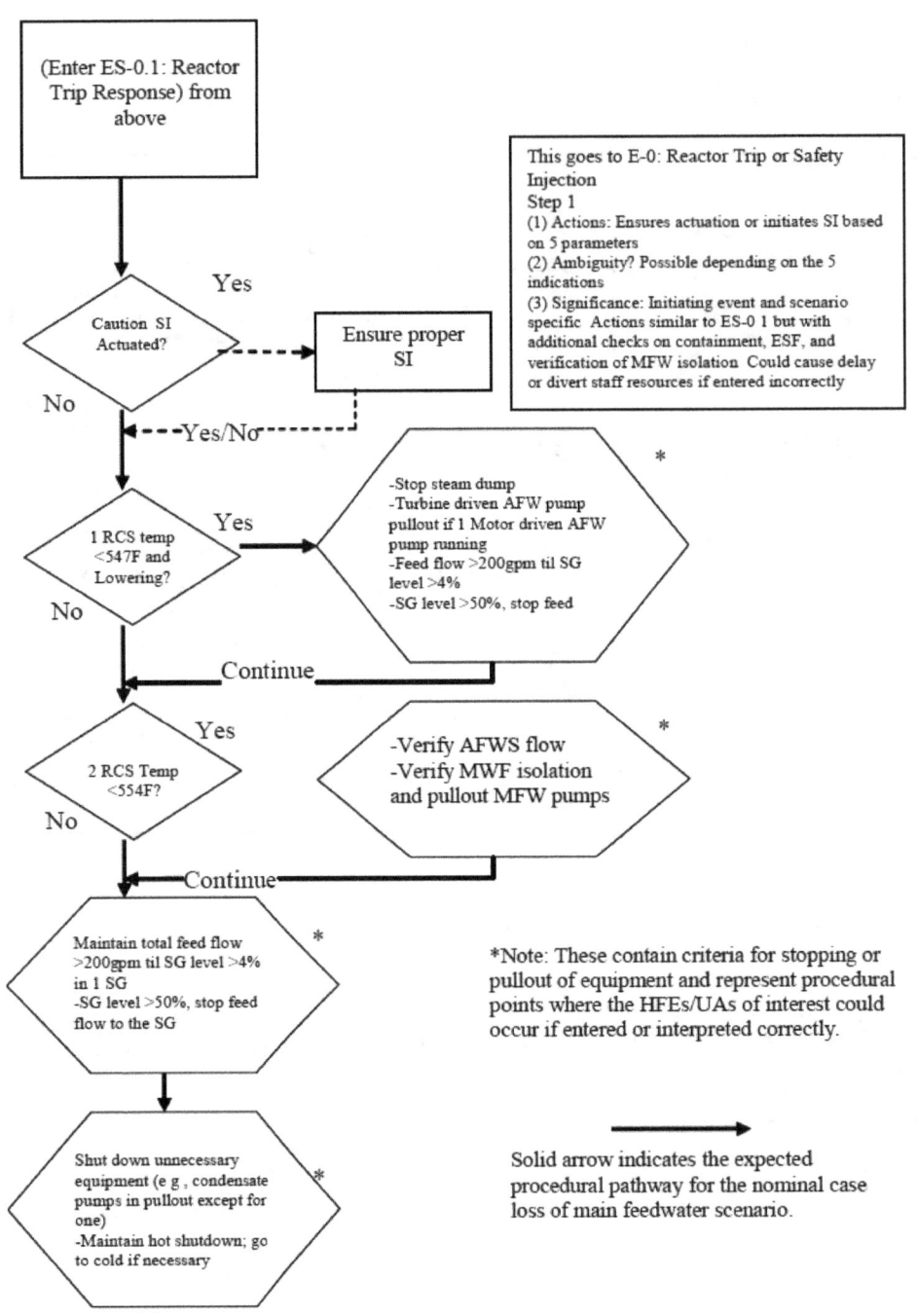

Figure A-1b. EOP Highlights Related to Loss of Main Feedwater Scenario (cont'd)

APPENDIX B

THEORETICAL BACKGROUND ON UNCERTAINTY
AND EXPERT ELICITATION

APPENDIX B

THEORETICAL BACKGROUND ON UNCERTAINTY
AND EXPERT ELICITATION

Sources of uncertainty. Uncertainties exist for many reasons including randomness (aleatory uncertainty) such as, inaccuracies in determination of the values of quantities and parameters (e.g., a probability value), high sensitivities in system performance to specific conditions, and omissions of important factors (e.g., a basic event, a specific PSF) from an analysis. They can also occur because of limits on our state of knowledge (epistemic uncertainty) such as incomplete knowledge regarding phenomena. Because of these multiple sources, discussions of uncertainty can easily become complex. See the text box, entitled "Aleatory and Epistemic Uncertainty" for a brief introduction.

Aleatory and Epistemic Uncertainty

It is helpful, in a very practical way, to characterize several classes of uncertainty, because careful thinking about the character of sources of uncertainty leads to a better representation of the associated pdfs. Pdfs for each class have very different shapes.(Siu et al. [Ref. 1]) Three classes of uncertainty are considered:

Deterministic Case—When there is no variability or there is no imperfect state-of-knowledge that leads to variability in the results:
- Aleatory uncertainty—When there is random variability in any of the factors that lead to variability in the results.
- Epistemic uncertainty—When the state of knowledge about the effects of specific factors is less than perfect.

To help understand these terms, a more operational point of view is that uncertainty is aleatory if:

- it is (or is modeled as) irreducible or
- the uncertainty is observable (i.e. repeated trials yield different results) or
- repeated trials of an idealized thought experiment will lead to a distribution of outcomes for the variable and thus this distribution is a measure of the aleatory uncertainties in the variable

The uncertainty is epistemic if:

- we are dealing with uncertainties in a deterministic variable whose true value is unknown or
- repeated trials of a thought experiment involving the variable will result in a single outcome, the true value of the variable, or
- it is reducible (at least in principle).

All of these sources of uncertainty can be portrayed in an analysis by means of a probability density function (pdf). This is a normalized function that portrays the relative likelihood that an uncertain variable will be observed within a particular interval.

The approach for the treatment of uncertainty implements the subjective framework for treating probabilities (Apostolakis [Ref. 2]). This approach provides the benefit of a clearer elicitation-based quantification process, in the following section. This benefit arises from the subjective framework's distinction between aleatory and epistemic uncertainties, which requires a careful examination of the factors contributing to uncertainty, resulting in a sharper definition of the issues being addressed during the elicitation. It provides an explicit way to tell the truth about what we know and what we don't know. Because of this and with thorough documentation, the analysis or elicitation can be well defended.

When a pdf is used to portray *epistemic* uncertainty it reflects a state of belief by an evaluator as described in the next section; the shape of the distribution tends to be a set of delta functions (two or more outcomes are possible and, once we run the right experiment, we'll know which describes the true state of the world). When a pdf is used to portray *aleatory* uncertainty it reflects the variability observed in repeated trials and usually has some kind of bell shape.

Bayesian Statistical Analysis. For risk assessment including HRA, the Bayesian approach offers great advantages because, for instance, all kinds of evidence are used and evaluating data with zero failures in N trials is straightforward. Most importantly, it gives meaning to using probability to describe epistemic uncertainty (Siu and Kelley [Ref. 3]). The basic idea is simple. From Bayes Theorem, the probability of two events A and E is:

$$P(A \wedge E) = P(A) * P(E|A) = P(E) * P(A|E)$$

which is a simple, uncontroversial statement of Bayes theorem. However, the Rev. Thomas Bayes [Ref. 4] went a step further. By rearranging as

$$P(A|E) = P(A) * P(E|A) / P(E)$$

and interpreting these terms as follows:

> $P(A|E)$ is the "posterior" probability of A, after collecting evidence E
> (say the result of an experiment)
> $P(A)$ is the "prior" probability of A (i.e., before the evidence is collected)
> $P(E|A)$ is the "likelihood" of the evidence (if the evidence can take on several values E_1,
> E_2,... the likelihood is the probability getting evidence E_i, given the prior)

This Bayesian switch (called inverse probability), lets us use something we know or can calculate [the likelihood of the evidence, $P(E|A)$] to determine the value of P(A given all the evidence at hand). It leads to a meaningful definition of subjective or state-of-knowledge probability. Details of the approach can be found in many standard sources (e.g., De Finetti [Ref. 5] and Jefferies [Ref. 6]).

With this background, what is meant by expert elicitation and how it is structured is addressed below.

Elicitation of Expert Evidence. How can we develop the uncertainty distributions that are needed for the analysis when so-called "objective" or "statistical" evidence (aleatory distributions built up through experience) is not available? Much of the following discussion is taken from four papers that catalog a wide range of the literature on this topic: Bley, Kaplan, and Johnson [Ref. 7], Budnitz et al. [Ref. 8], Forester et al. [Ref. 9] and Siu and Kelly [Ref. 3]. That literature originates among several independent disciplines, including operations research, psychology, risk assessment, and others.

It is important to address some of the criticisms and concerns that are often directed at analyses that rely, in part, on the opinions of experts. Some people believe that elicitation of expert opinion is just "guessing." Others worry that biases make human assessment of probability difficult or impossible.

When considering these concerns, it is important to recognize that the purpose of HRA and PRA is to support decision-making, that decisions are always made under uncertainty, and that decisions are going to be made, with or without the support of analysis. Therefore, analysts sometimes argue that when objective[12] or statistical evidence is available, they use it; however, when it is not, expert opinion is better than nothing.

This answer does not satisfy the critics, however. They recognize that expert opinion may be all we have, but they are not convinced that we are using that information with the proper caution. They suspect that we believe and trust the experts' opinions far more than we should, and they cite remembered cases in which the greatest experts in a field pronounced opinions that were subsequently proven to be totally wrong.

The questions of how to use expert opinion and of how to combine the opinions of different experts have generated much literature and much debate, and there remains disagreements even today. The text box "The Traditional Expert Opinion Approaches" explains two theoretical approaches for using expert opinion. The approaches differ in their characterization of uncertainty, but suffer from the same difficulties:

- Experts are often asked for opinions about parameters they do not use everyday.

- There is no demonstrably correct or best way to combine the estimates (with algebraic averaging, high outliers can dominate results; with geometric averaging low outliers can dominate).

- There are no built-in controls for bias, inconsistency in knowledge base, variability in interpreting the question.

However, there is a way to bypass many of the pitfalls of expert opinion. It begins by reframing the problem, and that begins by observing that which makes experts "expert" is not their opinions but their knowledge, experience, experiments, etc. — in short, their *evidence*. Therefore, instead of asking the experts for their opinions, we ask them for their evidence.

---

[12]   The notion of objectivity and objective information is often more hope than fact. Even when data exist, they come from very specific conditions and environments that may not apply to the case at hand. The same goes for experimental results and careful calculations; the real-world environments and upset conditions seldom exactly match the pristine experimental conditions or calculational assumptions. Rather than better-than-nothing, the application of careful engineering judgment to assess the applicability of "objective" information and to adjust it for real-world conditions is always necessary.

Kaplan suggested this reframing and called it the "expert information" approach [Ref. 7]. It has since been adopted in slight variation by a number of others [e.g., Refs. 8 and 9].

In the expert information approach, we do not ask the experts directly for their opinion about an elemental parameter, λ, that is in question (in our case, the HEP). Instead, we ask them what experience and information they have that are relevant to the value of λ. A facilitator then leads the group in combining the different kinds of information and evidence into a consensus state-of-knowledge curve.

The motivation behind this approach is the thought that, while the experts from whom we are eliciting information presumably have much knowledge in their particular domains, they usually are not trained or experienced in the use of probability as a language with which to express a state of confidence or state of knowledge. The latter subject is more the expertise of the facilitator/analyst. In addition, issues of bias and honesty (French [Ref. 17], Van Steen 1988a [Ref. 14]) conscious or unconscious, arise when the experts are asked to give λ_j, or $p_j(\lambda)$. These issues are bypassed if we go to the root, so to speak, and ask the experts for their evidence rather than for their opinions.

The expert information approach can thus be viewed as a formulation #3, which stands (see Table B-1) as a natural progression in relation to formulations #1 and #2 above and is outlined in the last text box below.

The Traditional Expert Opinion Approaches
(Bley et al. [Ref. 7])

Suppose that, in our PR&PP model of a specific design, we have a certain "elemental parameter," λ. This parameter is elemental in that it is not expressed in terms of any other parameters in the model. It itself must be entered as a basic input parameter. So, for the analysis, the question that must be answered is, "What is the numerical value of the parameter λ?" We seek to answer this question by putting it to the experts.

In the usual expert opinion approach [e.g., Refs 10–16] to eliciting and combining expert opinion, the problem is formulated in one of the following two ways:

1. Let $\lambda_1...\lambda_n$ be the point estimates of this parameter given by n different experts. Let $E_p = \{\lambda_i\}$ stand for this set of n point estimates. What shall we take as $p(\lambda|E_P)$, the probability curve representing our state of knowledge about λ, given the evidence E_p?

2. Let $p_j(\lambda)$ be the probability curve expressing the full state of knowledge of the jth expert, and let $E_f = \{p_j(\lambda)\}$ stand for this set of n probability curves. How do we combine these into $p(\lambda|E_f)$, the curve expressing our state of knowledge about λ, given E_f?

In formulation 1, the individual estimates, λ_i, are regarded as "data." The problem then becomes structurally identical with an everyday problem of experimental science; namely, given n different measurements of the quantity λ_i, what is our final state of confidence about its true value? Formulation 2 attempts to get more information out of each expert and thus is a bit of a stretch on the everyday problem. Nevertheless, in both formulations, the approach is that of an experimenter; we put the question, "How much is λ?" to nature (the experts), and we regard the answers, λ_i or $p_i(\lambda)$, as the resulting data. The problem for the analyst then centers around the determination of "weights," w_i, of some kind [Ref. 16] with which to combine the several answers.

B-4

Table B-1. Comparison of Three Expert Elicitation Formulations

Formulation	Question	Form of Answer
1	What is your best estimate for λ?	λ_j
2	What is your state of confidence about the value of λ?	$p_j(\lambda)$
3	What evidence and information do you have relevant to the value of λ?	E_j

---

The Expert Evidence Approach

Let E_j be the total body of evidence given by the jth expert. E_j should then constitute everything the expert knows that is relevant to λ. It is what makes the expert an expert. The idea of the expert information approach therefore is to elicit from the experts what they know best, E_j, and let the facilitator/analyst take the lead in codifying this knowledge into the form required by the HRA/PRA.

To do this, the facilitator/analyst first establishes a "total," or "consensus," body of evidence, E_T. This is the body of evidence that all of the experts agree to after reviewing and discussing each other's E_j. The analyst then guides the group of experts to a consensus probability curve, $p_c(l|E_T)$, on which they all agree.

References

[1] Siu, N.O., S. Malik, D. Bessette, and H. Woods, "Treating Aleatory and Epistemic Uncertainties in Analyses of Pressurized Thermal Shock," in *Probabilistic Safety Assessment and Management: Proceedings of the 5th International Conference on Probabilistic Safety Assessment and Management (PSAM 5), Osaka, Japan, November 27 – December 1, 2000*, Vol. 1/4, pp. 377–382, S. Kondo, and K. Furuta, Eds., Universal Academy Press, Inc., Tokyo, Japan, 2000.

[2] Apostolakis, G.E., "The Concept of Probability in Safety Assessments of Technological Systems," *Science*, 250(4986):1359–1364, American Association for the Advancement of Science, Washington, DC, December 7, 1990.

[3] Siu, N.O., and D.L. Kelly, "Bayesian Parameter Estimation in Probabilistic Risk Assessment," *Reliability Engineering & System Safety*, 62(1):89–116, Elsevier, Amsterdam, The Netherlands, October 1998.

[4] Bayes, T., "An Essay Towards Solving a Problem in the Doctrine of Chances," *Phil. Trans. Roy. Soc.*, 53:370–418 and 54:296–325, Royal Society Publishing, reprinted in *Biometrika*, 45:296–315, London, England, 1958.

[5] De Finetti, B., *Theory of Probability: A Critical Introductory Treatment*, Translated by A. Machi and A. Smith, John Wiley & Sons, New York, NY, Vol. 1, 1974, and Vol. 2, 1975.

[6] Jeffreys, H., *Theory of Probability*, Third Edition, Oxford University Press, New York, NY, 1961.

[7] Bley, D.C., S. Kaplan, and D.H. Johnson, "The Strengths and Limitations of PSA: Where We Stand," *Reliability Engineering & System Safety*, Elsevier, Amsterdam, The Netherlands, 38(1/2):326, 1992.

[8] Budnitz, R.J., G.E. Apostolakis, D.M. Boore, L.S. Cluff, K.J. Coppersmith, C.A. Cornell, and P.A. Morris, "Use of Technical Expert Panels: Applications to Probabilistic Seismic Hazard Analysis," *Risk Analysis*, 18(4):463–469, 1998.

[9] Forester, J., D.C. Bley, S. Cooper, E. Lois, N. Siu, A. Kolaczkowski, and J. Wreathall, "Expert Elicitation Approach for Performing ATHEANA Quantification," *Reliability Engineering & System Safety*, Elsevier, Amsterdam, The Netherlands, 83:207–222, 2004.

[10] Apostolakis, G.E., "On the Use of Judgment in Probabilistic Risk Analysis," *Nuclear Engineering and Design*, Elsevier, Amsterdam, The Netherlands, 93:161–166, 1986.

[11] Cooke, R.M., *Experts in Uncertainty: Opinion and Subjective Probability in Science*, Oxford University Press, New York, NY, 1991.

[12] Mosleh, A.M., and G.E. Apostolakis, "Models for the Use of Expert Opinions: Low-Probability/High-Consequence Risk Analysis," pp. 107–124, Plenum Press, New York, NY, 1984.

[13] Mosleh, A.M., V.M. Bier, and G.E. Apostolakis, "A Critique of Current Practice for the Use of Expert Opinions in Probabilistic Risk Assessment," *Reliability Engineering & System Safety*, Elsevier, Amsterdam, The Netherlands, 20:63–85, 1988.

[14] Van Steen, J.F.J., and P.D. Oortman Gerlings, *Expert Opinion in Probabilistic Safety Assessment*, Vol. 1: "Literature Study, Netherlands Organization for Applied Scientific Research," Delft University of Technology, Delft, The Netherlands, 1988.

[15] Van Steen, J.F.J., "Expert Opinion in Probabilistic Safety Assessment." in *10th Advances in Reliability Technology Symposium*, G.P. Libberton, Ed., pp. 13–26, Elsevier Applied Science, London, England, 1988.

[16] von Winterfeldt, D., E.-J. Bonano, R. Keeney, and S.C. Hora, "The Formal Elicitation of Expert Judgments for Performance Assessment of HLW Repositories," *Probabilistic Safety Assessment and Management*, G.E. Apostolakis, Ed., Elsevier Scientific Publishing, New York, NY, pp. 653–658, 1991.

[17] French, S., "Group Consensus Probability Distribution: a Critical Survey," in *Bayesian Statistics*, Vol. 2, pp 183–202, J.M. Bernardo, M.H. DeGroot, D.V. Lindley, and A.F.M. Smith (Eds), Elsevier Science Publishers, New York, NY, 1985.